The Promise of
Multispecies Justice

The Promise of Multispecies Justice

Edited by Sophie Chao,
Karin Bolender,
and Eben Kirksey

DUKE UNIVERSITY PRESS
Durham and London 2022

© 2022 DUKE UNIVERSITY PRESS
All rights reserved
Printed and bound by CPI Group (UK) Ltd, Croydon, CR0 4YY
Designed by A. Mattson Gallagher
Typeset in Garamond Premier Pro and Avenir LT Std
by Copperline Book Services

Library of Congress Cataloging-in-Publication Data
Names: Chao, Sophie, editor. | Bolender, Karin, [date] editor. |
Kirksey, Eben, [date] editor.
Title: The promise of multispecies justice / edited by
Sophie Chao, Karin Bolender, and Eben Kirksey.
Description: Durham : Duke University Press, 2022. |
Includes bibliographical references and index.
Identifiers: LCCN 2022002040 (print)
LCCN 2022002041 (ebook)
ISBN 9781478016250 (hardcover)
ISBN 9781478018896 (paperback)
ISBN 9781478023524 (ebook)
Subjects: LCSH: Environmental justice. | Environmental
ethics. | Environmental responsibility. | Restoration
ecology. | Physical anthropology. | Ecology—Moral and
ethical aspects. | BISAC: SOCIAL SCIENCE / Anthropology /
Cultural & Social | SCIENCE / Environmental Science (see
also Chemistry / Environmental)
Classification: LCC GE220.P766 2022 (print) |
LCC GE220 (ebook) | DDC 179/.1—dc23/eng20220627
LC record available at https://lccn.loc.gov/2022002040
LC ebook record available at https://lccn.loc.gov/2022002041

Cover art: Barbed wire and plant. Photograph
© 999041 / Dreamstime.com.

This book is dedicated to Deborah Bird Rose,
whose work continues to shimmer in the afterlife.

Contents

Acknowledgments ix

Introduction 1
Who Benefits from Multispecies Justice?
Eben Kirksey and Sophie Chao

Glossary 23
Species of Justice
Sophie Chao and Eben Kirksey

Blessing 29
Thanksgiving in the Plantationocene
Craig Santos Perez

1 **Spectral Justice** 33
Radhika Govindrajan

2 **Rights of the Amazon in Cosmopolitical Worlds** 53
Kristina Lyons

3 **"We Are Not Pests"** 77
Alyssa Paredes

4	**Prison Gardens and Growing Abolition** 103	
	Elizabeth Lara	
5	**Justice at the Ends of Worlds** 125	
	Michael Marder	
6	*from* the micronesian kingfisher 139	
	Craig Santos Perez	
7	**Rodent Trapping and the Just Possible** 157	
	Jia Hui Lee	
8	**Inscribing the Interspecies Gap** 179	
	M. L. Clark	
9	**Nuclear Waste and Relational Accountability in Indian Country** 185	
	Noriko Ishiyama and Kim TallBear	
10	**Multispecies Mediations in a Post-Extractive Zone** 205	
	Zsuzsanna Ihar	

Closing   227
Th S xth M ss Ext nct n
Craig Santos Perez

Afterword   229
Fugitive Jurisdictions
Karin Bolender, Sophie Chao, and Eben Kirksey

Bibliography   239
Contributors   273
Index   277

Acknowledgments

We acknowledge the peoples of the Kulin, Kalapuya, Eora, and Darug nations, for allowing us to work and live on their unceded lands while compiling this book collection. We acknowledge that genocidal and colonial processes in Australia, North America, Melanesia, and elsewhere shaped and constrained our knowledges. We are grateful to the newly recognized essential workers of 2020, who supplied us with food, medical supplies, and other life-sustaining support during the pandemic.

This volume came together in distinct ways both because and in spite of the tumultuous unfoldings of recent years. Beginning in the spring of 2020, a steadfast crew of scholars and artists came together regularly in a dedicated online space, across a dizzying array of time zones and living situations, to present and discuss the ideas and texts presented here. The camaraderie, provocations, and sense of common purpose cultivated in that space created a steadying force in times of unforeseen global upheaval and mind-bending inflection. We are grateful for all the gifts brought to bear on this volume by participants in that MSJ (Multispecies Justice) crew and visitors, who shared their time and insights in the forms of chapters in progress, artworks, and conversations. Thank you to Marisol de la Cadena, Kuang-Yi Ku, Tessa Laird, Michele Lobo, Celia Lowe, Jan-Maarten Luursema, Santiago Martínez Medina, Gina Moore, Natasha Myers, and Anna Tsing.

Participants in the Promise of Multispecies Justice series of online talks held in 2021 offered further depth and breadth to the ideas presented in this book. We thank in particular our guest presenters, Danielle Celermajer, Ursula Heise, and Macarena Gómez-Barris, and our guest discussants, Alex Blanchette, Carla Freccero, Evan Mwangi, Nicholas Shapiro, and Christine Winter, and also our virtual auditors and interlocutors who tuned in from around the world.

We extend heartfelt gratitude to our editor at Duke University Press, Ken Wissoker, for his precious mentorship and wise guidance in helping bring this book to fruition, and to Ryan Kendall for supporting us throughout the production process. Immense gratitude also goes to the two anonymous reviewers of this manuscript for their copious and constructive feedback.

The leadership of Emma Kowal and Fethi Mansouri of the Alfred Deakin Institute created a generative environment for us to nurture the initial seeds of ideas for this project and bring it to life. Financial support for this book was received from the Australian Research Council under a Discovery Project Grant (Ref: DP200102763). Luke Cuttance, Pauline Seet, and Giles Campbell-Wright gave us strategic support throughout. Eben Kirksey is particularly grateful for the intellectual community of Deakin University—especially Vanessa Lemm, Miguel Vatter, David Turnbull, Joe Graffam, Chris Mayes, and David Giles—for companionship and generative conversations amidst the contingencies of lockdown. Inspiration on justice came from Julieta Aranda, while Leon Aranda staged creative disruptions.

As a postdoctoral research fellow at the University of Sydney's Department of History, Sophie Chao focused on bringing this book to fruition. Particular gratitude goes to the Janet Dora Hines Endowment and to Warwick Anderson, Keith Dobney, Annamarie Jagose, and members of the School of Philosophical and Historical Inquiry for this unique opportunity. At the University of Sydney, reading groups and symposiums held by the Sydney Environment Institute Multispecies Justice Collective have offered vital interdisciplinary insights into theories and practices of justice. Particular thanks go to Danielle Celermajer, Anna-Katharina Laboissiere, Astrida Neimanis, David Schlosberg, Blanche Verlie, Dinesh Wadiwel, Christine Winter, and Katie Woolaston.

Colleagues from a wide range of fields and disciplines provided critical feedback on the arguments and ideas presented in this book at various guest seminars and conferences. These include presentations that Sophie Chao delivered at the University of Manchester, New York University, Central University of Karnataka, Cornell University, University of California Berkeley, London School of Economics, Australian Academy of the Humanities, Kunsthistorisches Institut in Florenz–Max-Planck-Institut, University of

Waikato, Asia Research Institute, and Yale University–National University of Singapore College, and Sydney Environment Institute. Chao thanks the organizers and the participants of these events for their incisive questions and comments. Last but by no means last, Chao thanks her family, Dominique Chao, Jacques Chao, and Emmanuel Chao, and her partner, Jacob Perrott, for their unfailing support and ongoing care.

Karin Bolender is grateful for the new paths this project opens toward investigating sensory and other more-than-human dimensions of possible justices via interwoven creative and critical forays. This project gained important insights from Anne Riley, T'uy't'tanat-Cease Wyss, JuPong Lin, and Laiwan; Eli Nixon and the wondrous Bloodtide workshop celebrants; Heather Barnett's international Slime Mould Collective, and slime mold Andi; angela rawlings and MOONLINE; Emily Eliza Scott; and Composting Feminisms. In the urgent quest to bring practical questions of how to do justice for all into contested (stay-at-)home places, Karin Bolender basks in a stream of inspiration from Annie Sprinkle and Beth Stephens, dance for plants, Laboratory for Aesthetics and Ecology, and Kultivator. On-the-ground companionship in the course of this project came from a vibrant community of ecological artists and curatorial agents (thanks to Agnese Cebere and Eugene Contemporary Art's *Common Ground*, and the R. A. W. Pulping Posse); furred, furless, feathered, and photosynthesizing friends and kin with keen noses for more-just assemblages; and the care, endurance, and ingenuity of Rolly and Sean Hart, holding it down on the home front and other swirling horizons.

Vital technical and logistical support for this project was offered by research assistants Marita Seaton and Rebecca Willow-Anne Hutton, proofreader Marc Herbst, and web designer Alex Reeves of Moonpool. We also thank artist and scholar Feifei Zhou for creating many of the artworks featured in this book.

Finally, we thank the contributors to this volume for their tireless commitment and generosity of self and spirit throughout our collaborative journey: M. L. Clark, Radhika Govindrajan, Zsuzsanna Dominika Ihar, Noriko Ishiyama, Elizabeth Lara, Jia Hui Lee, Kristina Lyons, Michael Marder, Alyssa Paredes, Craig Santos Perez, and Kim TallBear.

With this book we hope to create an opening for future research and advocacy on multiple fronts that brings the promise of justice into contact with historical reality.

Introduction
Who Benefits from Multispecies Justice?

Eben Kirksey and Sophie Chao

EARLY WORK IN MULTISPECIES STUDIES described how symbiotic associations and the mingling of creative agents generated emergent ecological communities. Multispecies ethnographers mobilized approaches from cultural anthropology to study plants, animals, microbes, and fungi whose lives and deaths are intertwined with human social worlds. Justice and injustice were part of the conversation since the beginning of the field, even though these concerns were ancillary to early texts rather than the central focus. Anthropologists, sociologists, geographers, and many other multispecies scholars have followed Susan Leigh Star who suggested it is more analytically interesting and more politically just to begin with the question "Cui bono?" (Who benefits?) than to simply celebrate the fact of human/nonhuman mingling.[1]

Sympathetic criticism of multispecies ethnography has led many scholars to explicitly focus on justice as well as political and ethical concerns. A decade ago, when the field was still finding its feet, Kim TallBear emphasized the importance of including Indigenous and queer standpoints as we reckon with the violence generated by Western cosmologies that have divided human and nonhuman realms.[2] Many writers have since abandoned the "nonhuman" altogether, since it implies the lack of something—like "nonwhite." As we conceptualize multispecies futures, it is important to keep the intergenerational legacies of colonialism and racism in mind. Intersectional political movements and research practices are starting to simultaneously address issues related to social justice and interspecies care.[3]

The phrase *multispecies justice* was introduced by Donna Haraway in 2008 with her foundational book, *When Species Meet*. As we continue to build on classic insights about how human existence is bound together with the lives of other entities, it is possible to address emergent intersectional concerns. Building on an influential and quickly growing body of work,[4] *The Promise of Multispecies Justice* addresses an array of intersecting questions: Who are the subjects of justice in our shared worlds? What is at stake when they are captured by juridical-legal systems and social movements? Who has claimed a monopoly over justice in the past, and in the present, and how might we contest their sense of propriety in the future?

Injustice (the lack of something) is often more tangible than justice (the supposed fullness or perfection of something).[5] In assembling this collection of essays, we searched for authors who were focused on situated struggles against ugly injustices while also attuned to beautiful creatures and multispecies communities that are sources of delight.[6] In a book entitled *On Beauty and Being Just*, Elaine Scarry observes that there does not appear to be a finite amount of space, inside the brain, given to beautiful things or just causes. Since ideals about beauty can be hegemonic, Scarry is particularly concerned with errors in aesthetic judgement. She recalls the disorienting shock when suddenly she notices a palm tree—a kind of plant she had not previously recognized as noteworthy—"arcing, arching, waving, cresting and breaking in the soft air, throwing the yellow sunlight up over itself and catching it on the other side."[7]

Writers in the multispecies tradition have worked to cultivate what Anna Tsing calls "arts of noticing."[8] This approach involves paying attention to charismatic forms of life—like plants, birds, and butterflies—as well as creatures that are often disregarded or actively targeted for destruction like mushrooms, bacteria, rodents, and beetles.[9] The idea is not just to celebrate the beautiful, like Scarry's palm tree, but also to notice uncanny presences, such as stray dogs and thorny plants in post-industrial Azerbaijan (see Ihar) or the carcass of a dismembered bull by the side of the road in the Indian Himalayas (see Govindrajan). Writers who practice these multispecies arts of noticing have begun to turn away from landscapes conventionally regarded as beautiful—

like protected parklands and conservation zones—and toward sites of abandonment and extraction, like toxic waste dumps and plantations.[10] Feral forms of life have become sources of wonder in the midst of dread.[11] *The Promise of Multispecies Justice* embraces these alternative aesthetic sensibilities.

This book is the result of coalitional thinking with gardeners, anti-racist activists, and Indigenous peoples, as well as the urban and rural poor in Asia, the Caucasus, Africa, and the Americas. During the height of the 2020 pandemic summer, we assembled an international and interdisciplinary crew in a Zoom room to share stories with glimmers of hope for peoples and creatures who are united across geographical divides by shared experiences of violence, humiliation, or abandonment. We gathered writers with expertise—in cultural anthropology, geography, philosophy, science fiction, poetry, and fine art—to track the contours of justice as it travelled from courtrooms to protest movements in the streets, from the abstract realms of high theory to the fleeting domain of ghosts and spirits.

Together, we explored intersectional alliances emerging among diverse peoples and species on planet Earth, and even the possibilities of transformative justice in extraterrestrial realms (see Clark). We found that dominant hierarchies of life and worth—placing humans above other species—were being subverted and resisted in diverse cultural and ecological milieus. In Micronesia, we were drawn to Indigenous practices for exalting the beauty of an endangered bird—a kingfisher with green, orange, and white feathers—while colonial enterprises continue to produce extinction (see Perez). Plants growing in the prisons of California offered us glimpses of beauty within a deeply flawed justice system, and they illustrated possibilities for the prison abolition movement to gain ground and grow (see Lara). In the Colombian Amazon, open violence was intensifying even as dreams proliferated about justice for "Nature" itself (see Lyons). In these settings, and many others, we collectively asked: Can we depart from particular grounds of possible flourishing to bring justice to other sites and scales? How does expanding the scope of justice beyond the human and the law invite new possibilities for decolonizing multispecies relations, and the concept and practice of justice itself?

Species of Justice

The idea of justice is often accompanied by qualifiers like social justice, restorative justice, and distributive justice. Naming forms of justice and injustice, Jessica Greenberg suggests, can produce a sense of action and agency. Creating a critical and activist semiotics, Greenberg argues, lets us experience the possibility of incremental hope and messy future action. "The opposite of justice is not injustice," she proposes, "but despair."[12] Rather than sink into a paralyzing politics of despair, this book identifies, defines, and indexes multiple species of justice (see Glossary). Our collection of essays reclaims modest forms of hope through accounts of transitional justice (see Lyons), multiworld justice (see Marder), small justices (see Ihar), generative justice (see Lee), and spectral justice (see Govindrajan). Taken together, these essays explore tactics for achieving multispecies justice in polymorphic situations where calculations are never perfect and are often open to reinterpretation.[13]

While the field of multispecies studies has largely moved past the environment as a framing concept for human agency and action, it is important to recognize the significant research about justice that is taking place in environmental studies broadly defined. The environmental justice movement emerged in the 1980s, as civil rights activists and scholars in the southern United States began protesting toxic waste dumping in marginalized communities.[14] The slow violence of environmental racism was damaging human bodies, even as overt violence and killings took place.[15] Early environmental justice struggles brought critical attention to the bias of government knowledge and industry practices.[16] Citizens impacted by environmental racism generated their own community-based participatory research initiatives.[17] Critical insights from these prior generations of activists and environmental scholars are more necessary than ever as we start to consider the promise of multispecies justice.

Our approach to multispecies justice is informed by Western continental philosophy and political theory related to rights and capabilities. Philosophers associated with the animal rights movement have drawn on the utilitarian tradition in ethical philosophy to deem actions right or wrong, depending on the extent to which they promote happiness or prevent pain. Justice can be accomplished, according to utilitarian thinkers, when people intuit the feelings of animals and become advocates for their welfare.[18] Yet, in speaking about possibilities of justice in multispecies worlds—in conversation with activists, biologists, nature lovers, environmental advocates, politicians, farmers, or philosophers—it is important to remain ever mindful of the creatures and communities being represented and who represents them.[19] Work in the field of multispecies studies always contains the risk of ventriloquism—the problem identified by Arjun Appadurai, as anthropologists attempt to speak "for" and speak "of" others.[20] Words often fail us when we attempt to do justice in multispecies realms (see Bolender, Chao, and Kirksey).

Capabilities-centered approaches to justice offer useful resources for getting beyond the problem of ventriloquism through the consideration of physical, psychological, emotional, and cognitive well-being of animals, as well as their social relationships and ecological interdependencies. Practices of "sympathetic imagining" across gulfs of species' differences, and creative engagements with scientific literature proposed by philosophers like Martha Nussbaum, offer fresh perspectives on the flourishing of animals as dignity-bearing subjects, agents, and world makers.[21]

Classically, theories of distributive justice were concerned with the equal distribution of resources and benefits as well as equal protection from scarcities and risk.[22] Struggles over distributing resources can tacitly reinforce hegemonic structures of power as claims are made on dominant institutions (see Lee).[23] Approaches to recognition justice instead acknowledge how different peoples or beings gain or lose standing as a result of structural, institutional, cultural, legal, and economic regimes as well as attendant hierarchies of worth.[24] Sociolegal scholars have expanded the conventional scope of this recognition by proposing "ecological vulnerability" as a theory of justice that positions the autonomous self within a larger relational framework of existence.[25] It is thus important to recognize how unequal vulnerabilities impact humans as well as species and ecosystems.

The contributors to this volume build on this expansive archive of Western continental philosophy and political theory, while remaining attentive to ways that justice has been twisted by colonialism, capitalism, and racism. On the margins of banana plantations of the Philippines, Alyssa Paredes explores justice at the limits of human empathy and sympathetic imagining beyond species lines. In a mountainous region of India, Radhika Govindrajan invites us to consider what recognition justice might entail when animals who suffer violent deaths continue to haunt the living with their ghostly presence. As the United States Department of Justice perpetuates the mass incarceration of Black and brown peoples, Elizabeth Lara's essay about prison gardens insists that "the work of abolition is also the work of multispecies justice."

The very principle of equality underlying distributive theories of justice (both in terms of personhood *before* the law and resources owed *to* subjects of the law) is breaking down in the context of racist police and judicial practices (see Lara), consequential species differences (see Paredes), and the ongoing theft of Indigenous sovereignty over lands, bodies, and animals (see Ishiyama and TallBear; Perez).[26] Procedural approaches to resolving injustice are, in many cases, failing to achieve substantive justice with concrete outcomes. As some aspire to produce transformative justice, what counts as the "greatest good," and for whom, often remains ambiguous (see Clark; Lyons).[27] Some essays in this collection explore situations of injustice where there is no clear consensus on what justice might mean in practice (see Govindrajan; Ihar).[28] Justice thus can be open-ended, elusive, or impossible.

The multiple species of justice that come together in this book offer what Claire Jean Kim calls a multioptic vision, or a way of seeing that takes different claims seriously without hastily granting undue importance to any particular approach.[29] Multioptic vision enables an onto-epistemological and methodological unflattening—a simultaneous engagement of multiple vantage points from which to engender new ways of seeing, imagining, and being in relation to other Others.[30] Imagining just futures through this intersectional framework demands that we reconsider what constitutes the threshold of (in)justice, who gets to determine it, and in whose interests.

Multispecies justice, at its core, is promissory as we share common dreams and tend to the creatures and communities that we love. "The promise," in the words of Sara Ahmed, "suggests happiness lies ahead of us, at least if we do the right thing."[31] But pronouns can be slippery as "we" imagine shared futures.[32] When we harbor hopes about desirable and undesirable species interactions, about intended and unintended consequences, about moments of justice to come, we bring together different communities of "us."[33] Multispecies justice emerges within fields of power where who is in the world, and whose world counts, is at stake.[34] Any project that aims to achieve justice in multispecies worlds should thus ask: justice for whom or what?

Rights of "Nature"

Courtrooms are perhaps the most recognizable public arena where people are pursuing something akin to multispecies justice within existing legal paradigms. New Rights of Nature laws are focused on inherent rights of ecosystems and species. Some legal cases within this framework have also furthered concerns related to social and environmental justice. For example, a 2018 court ruling in Ecuador that shut down the Río Blanco mine represents an important victory at the intersection of human social worlds and the lifeworlds of multiple species. This particular case brought together the concerns of local communities, municipal leaders, Indigenous Quechua, and environmental activists who speak for watersheds, wetlands, and páramo ecosystems.

Multispecies justice, for human and ecological communities, is often a provisional achievement. The ruling that closed the Río Blanco mine is just a temporary legal victory that might still be overturned with future appeals by the mining company. Even though the Ecuadorian government has exercised considerable "control over the meanings and the uses of 'rights' for nature," in the words of Erin Fitz-Henry, this particular case illustrates that the interests of the authorities can sometimes align with broad political coalitions that are pushing for justice on behalf of multiple species.[35] The Río Blanco ruling demonstrates that justice for people can also be attuned to multispecies relations.[36]

As Rights of Nature laws proliferate in jurisdictions around the world, ethnographers are just starting to study exclusions and conflicts that are reverberating beyond the walls of courtrooms. Kristina Lyons describes a particularly acute site of ongoing trouble in Colombia's Amazon as leaders of social and environmental movements reckon with a ruling from the distant center of power in Bogotá. Following a 2016 peace accord, transitional justice initiatives in Colombia have resulted in some important criminal prosecutions, but at the same time the targeted killing of environmental advocates has intensified.

As conservation efforts in Colombia become increasingly militarized, Lyons suggests that people involved in ongoing disagreements should aspire to have "better conflicts" as they try to resolve competing social and environmental problems. While many call for peace, Lyons builds on the work of Isabelle Stengers to propose "new modalities of warfare" in fighting for forests and other loved communities of life.[37] As other ethnographers approach these new modes of warfare in theory and praxis, it is critical to consider how antagonism might help us understand the democratic possibilities in multispecies milieus where conflict is *sustained*, rather than erased.[38] Antagonism can produce consequential shifts in the order of things—not a final peace, but ruptures in the established order that produce opportunities for new collaborations, alliances, and worlds.

While fighting for justice in legal and symbolic realms, while advocating for forms of life that we find beautiful and necessary, it is critical to remain mindful of exclusionary languages and logics.[39] Feminist theorists of science and society have already drawn attention to "implicated actants"—the animals, plants, species, and ecosystems—who are "silenced or only discursively present, constructed by others for their own purposes" in legal rulings and some environmental campaigns.[40] While government officials, lawyers, and some activists rally to defend Nature, it is important to return to a question that Donna Haraway posed more than twenty years ago: "What counts as nature, for whom, and at what cost?"[41]

Beyond the Limits of the Human

As legal scholars and philosophers seek to expand the category of rights-bearing subjects—to identify animals, plants, rivers, and ecosystems as legal persons—it is important to remember that not all people are treated as fully human before the law. Alexander Weheliye, among others, has described how unequal power structures determine "which humans can lay claim to full human status and which humans cannot."[42] Colonial constructs of race and nature haunt the cultural politics of identity as terrains of power operate within and across species lines.[43] As a result, some species and kinds of people continue to be vulnerable to premature death, or even targeted for extermination.[44] As open violence is directed at people on account of their blackness, brownness, and queerness, slower forms of violence are incrementally and often imperceptibly disabling and debilitating vulnerable peoples and species.[45]

In approaching the promise of justice, we—that is, Eben Kirksey and Sophie Chao—bring profound feelings of despair from long-term ethnographic research among Indigenous peoples who are living through conditions of genocide. Both of us work in the Indonesian-occupied region of West Papua, where a Black liberation movement is trying to achieve recognition on the international stage. Human rights laws are failing in West Papua. Dreams from the mid-twentieth century, about universal justice for humankind, are foundering here and many other parts of the world. The Indonesian army has systematically killed and tortured Indigenous Papuans in a pattern that has continued, largely uninterrupted, since their initial invasion in 1961.[46] One particular incident, where upward of 139 Indigenous Papuans were massacred, was witnessed by Kirksey in 1998. The rules of law were never applied to this particular case, nor to countless other state killings.

The conflict in West Papua over who counts as human intensified in 2018 after Indonesian social media influencers posted a picture of a Papuan leader alongside a gorilla. Anti-racist protests erupted a year later when a group of Papuan students were physically attacked and verbally abused—called monkeys, dogs, and pigs by members of an Indonesian militia. Some Papuan activists and Indonesian allies donned full-body primate costumes, masks, and headdresses as they challenged the imagery that renders some people subhuman, killable, and disregarded before the law.[47] Banners, placards, and slogans deployed during protests across Indonesia read: "We are not monkeys!" "The monkeys take to the streets!" and importantly "Monkeys stand united against the colonizers!" Meanwhile, a flurry of monkey-themed hashtags shared by Papuans and Indonesian supporters went viral on Twitter, Instagram, and Facebook. They included #PapuaBukanMonyet ("Papuans are not monkeys") and #KamiJugaManusia ("We are also humans") but also #SayaMonyet ("I am monkey").

Proximity to animals can be deadly—even if this closeness is just the result of a visual juxtaposition or a racial slur. The case of West Papua speaks to the experience of many different kinds of people—including disabled, queer, brown, Black, Indigenous, or minority religious communities—who continue to bear the burden of imposed comparisons to the nonhuman.[48] The symbolic violence of these comparisons can quickly give way to actual violence. People become vulnerable to incarceration and torture—or even become killable—when they are singled out, as Claire Kim notes, for "how animal they are—and, how human they are not."[49] Even still, proximity to animals can also be affirmative. The surprising identifications made by Papuans and their allies—"I am monkey" and "Monkeys stand united against the colonizers!"—speak to the radical forms of interspecies alliances that are emerging here and in other parts of the world.

Activists in West Papua and elsewhere are pushing back against ways of speaking and thinking that prioritize some lives over others—they are disrupting value-laden categories that separate the human from the nonhuman. By embracing the figure of the monkey, these Indigenous intellectuals are rejecting the paired logics of racialization and animalization, even as they oppose ongoing processes of colonialism by lighter-skinned Indonesian settlers. They are refusing to denigrate the animal, while agitating against destructive forces that threaten their own lives. This refusal resonates with writings by Bénédicte Boisseron, who has lately begun to explore the forms of solidarity and sociality that have emerged for peoples and animals who have experienced "entangled forms of oppression."[50]

Interspecies Intersectionality

Multispecies justice involves seizing opportunities for intersectional political praxis across species lines. Intersectionality has classically been used as an analytical framework to understand how social and political identities combine to create different lived experiences of discrimination and privilege.[51] Ideas about intersectional inequalities involving humans and other species have recently been explored in the context of a patchy Anthropocene, where concerns are simultaneously people-focused and also engaged with multispecies relations.[52] In an allied spirit, we are interested in patches of justice amidst uneven conditions of livability.

Political ecologists and economists have already identified intersectional connections between social justice principles and transnational environmental concerns.[53] Juan Martinez-Alier's account in *The Environmentalism of the Poor* describes how small-scale farmers and Indigenous peoples began to frame their political struggles in environmental terms in the 1980s. While environmental justice was a concern of minority groups in the United States, Martinez-Alier contends that these issues are important to "the *majority* of humankind, those who occupy relatively little environmental space, who have managed sustainable agroforestal and agricultural systems."[54] The Environmental Justice Atlas, a participatory mapping project started by Martinez-Alier in 2016, currently encompasses over 3,400 local social movements that are opposing the frontier logic of global capitalism in diverse corners of the world.[55]

Mapping does not necessarily or automatically translate into coalitional action. As thousands of environmental advocates each fight beautiful and necessary struggles on their own local patches of ground—in places like West Papua, the Amazon, and the contaminated lands of Azerbaijan—there is the potential of competition for attention on the international stage. Creating and sustaining effective alliances has been one of the central challenges of intersectional struggles, ever since Kimberle Crenshaw introduced the idea of intersectionality—questioning why white women had not created enduring coalitions with Black women.[56] Competitive struggles for justice can nonetheless transform into coalitional movements, as contingent and nonnecessary links are built and maintained.[57] Collective work—across different patches of justice—has enabled us to do some of this work through conversations with anthropologists, artists, poets, and authors of fiction. Together we have also approached the challenging work of pushing beyond "the environment" to understand the possibilities and perils of sustaining intersectional alliances across species lines.

In ecological communities filled with predators and prey, hosts and parasites—worlds where hostility and hostages are embedded at the very heart of *hospitality*—alliances can be fleeting as interests align, only to unravel again.[58] Even still, "symbiotic agreements" can emerge among interdependent subjects as former enemies become allies.[59] According to Isabelle Stengers, peace does not exist within ecological communities. Instead of longing for peace, she argues for the necessity of ongoing battles in sustaining conditions for life on earth. This "cosmopolitical proposal" demands compromise with destructive species and hostile peoples who challenge our fundamental ethics, values, and understanding of the cosmos, or even threaten to destroy our world.[60]

Since many political theorists suggest that subjects are never entirely self-aware, or rational, the idea of intersectionality should travel well into cosmopolitical and multispecies worlds to account for how different kinds of entities move each other as their interests align.[61] While the idea of agency implies fully present, individuated, and autonomous subjects, the framework of intersectionality points to possibilities of identification among differently embodied, situated, and entangled beings.[62] Intersectionality points to the hopeful edge of interspecies political projects—where new articulations might produce new grounds for possible flourishing.

Failures in identification nonetheless abound. In the Philippines, a political movement has emerged on the margins of banana plantations, where people assert a strong rhetorical divide between humans and nonhumans, even as their livelihoods depend on animals and plants that share their conditions of precarity. Alyssa Paredes describes how Filipino activists have united around a rallying cry—"We Are Not Pests"—while demanding justice after being sprayed with toxic fungicides. In aspiring toward an alternative vision of multispecies justice, Paredes proposes a new slogan for confronting the violence of contemporary plantations: "We, Too, Are Pests!" This alternative rallying cry contains echoes of the counterintuitive slogans from West Papua that point to possibilities for interspecies alliances even amidst conditions of genocide. These ideas also resonate with the deep archive of writing in the tradition of Black ecology that shows how people and animals, linked by the inhuman logics and violence of slavery, share, in the words of Joshua Bennett, "the desire for a world without cages or chains."[63]

Unflattening

Amid ongoing struggles over water and gas pipelines in North America, the militarization of Indigenous lands in South America, as well as megamines and plantations in all parts of the world, Macarena Gómez-Barris observes that "indigenous and multispecies autonomy are increasingly in peril." She calls for more innovative thinking and writing "against the binary of the human and non-human."[64] However, some creative multispecies writing has attracted criticism for evacuating the diversity and complexity of human lifeworlds.[65] Theoretical and rhetorical gestures that result in a "flattened multispecies ontology—where difference among and between forms of life is obscured"—have also been critiqued by Janae Davis and her colleagues. These scholars suggest that sustained engagements with racial justice struggles might inform new conceptual and empirical research about "multispecies assemblages that lead out of socioecological crises toward better futures."[66]

An argument for flattening ontologies is articulated in Jane Bennett's *Vibrant Matter*, the book about "thing power." Reveling in the liveliness and agency of all matter, Bennett blurs distinctions between persons and things such as a dead rat, oak pollen, a plastic glove, and a bottle cap. Collapsing the ontological divide between people and things, she admits, "will not solve the problem of human exploitation or oppression." Instead, she hopes to solve other problems by promoting ethical behavior in the realm of vital materialism. "In a knotted world of vibrant matter, to harm one section of the web may very well be to harm oneself," Bennett suggests. "Such an enlightened or expanded notion of self-interest *is good for humans*."[67]

Many multispecies ethnographers and allied anthropologists have consistently pushed back against this flat conception of ontology to scrutinize the interplay of distinct agents and entities amidst flows of power. While Bennett attempts to flatten the distinction between life and nonlife, Elizabeth Povinelli asks: "Are we simultaneously extending the qualities and dynamics of one form that we believe existence takes (Life) onto the qualities and dynamics of all forms of existence?"[68] Rather than focus on the self-interest of humans, multispecies ethnographers have considered how creatures once limited to the realm of zoe—bare life, which is killable—have begun to appear alongside humans in the realm of bios, with biographical and political lives.[69]

As some scholars call for abandoning the species concept, we contend that this idea remains useful to the work of ontological unflattening in more-than-human worlds.[70] Humans are not alone in the world in recognizing a plurality of species, beings, and kinds. Other creatures—like frogs, plants, and even bacteria—also engage in their own practices of classification, recognition, and differentiation as they live together in shared multispecies milieus.[71] In the words of the philosopher of science John Dupré, species can be recognized with a "promiscuous realism" that stabilizes important aspects of reality amidst radical ontological pluralism and general metaphysical disorder.[72]

Anthropologists have lately started to push the species concept beyond the realm of biology, while holding on to consequential phenomenological differences between domains of the living and the nonliving.[73] Within the nonliving realm, chemists use the idea of species to denote identical molecular entities that undergo dynamic changes through processes like combustion, decay, and sublimation. Multispecies ethnographers and scholars in allied fields have started to use conceptual and technical tools for grappling with this material reality to ask: "How are molecular frictions, catalytic dynamics, forms of not-Life, and other-than-life reconfiguring our conditions of knowing, being, and sociality?"[74]

The Promise of Multispecies Justice showcases the work of scholars, activists, and poets who use precise and promiscuous languages as they name creatures and communities suffering ongoing harm as life becomes nonlife on a planetary scale. Several essays in this collection focus renewed attention on the interplay of chemical and biological species, in situations where people are working to actualize various kinds of justice. In the windswept desert landscape of Baku, Azerbaijan, displaced peoples, plants, and dogs have found distributed possibilities for justice in abandoned oil-contaminated lands (see Ihar). As toxic chemical plumes drift from banana plantations—harming people, plants, and animals in the Philippines—community activists are trying to achieve justice in courtrooms where they are required to identify and isolate "Bad Actor Chemicals" (see Paredes). In these situations, ethnographers are developing strong forms of knowledge—with the potential to disrupt hegemonic relations of power—while deriving theoretical insights about dynamic molecular and multispecies intra-actions.[75]

Struggles for multispecies justice—especially in arenas adjacent to biology and chemistry—risk being captured by technoscientific modes of knowledge production. Efforts to build universalized knowledge and justice together can produce both epistemological and ontological forms of exclusion.[76] Beyond tangible realms, Ruha Benjamin warns that "with increasing attention to the possibility of forging multispecies justice ... there has been far less attention to immaterial actants such as those inhabiting the ancestral landscapes."[77] This is why this book reckons with the haunting presence of souls and spirits alongside abiotic, elemental, and molecular entities (see Govindrajan; see Paredes).

Native American scholars are blending approaches from the social and natural sciences to understand human entanglements with animals, plants, and other entities like stones and water.[78] While making a soft refusal of the terms *spiritual* and *species*, our contributors Noriko Ishiyama and Kim TallBear invite us to think about how biological and social relations might contribute to an "idea of co-becoming that refuses the nature/culture divide." Other authors in this collection remain committed to the species concept as a valuable tool for making sense of the ebb and flow of agency in multispecies worlds. Memorizing and reciting a scientific name—like *Halcyon cinnamomina cinnamomina* for the Micronesian Kingfisher—can produce curiosity and wonder about life at the edge of extinction. Craig Santos Perez's poem about this endangered bird species also shows how Indigenous names—like sihek—might animate lively futures.

By bringing together authors who represent different ontological standpoints and political visions, we have built on the work of Nick Sousanis, whose genre-bending comic book, *Unflattening*, illustrates pathways from a one-dimensional flatland toward multidimensional possibilities.[79] We recognize that differential relations to power mean that not all peoples or species can equally access the possibilities contained in the future itself. Within future-oriented imaginaries shaped by science, technology, and justice, Ben Hurlbut observes, "Particular conceptions of progress, of the human person, and of the good are engaged, displacing others."[80]

Reclaiming the Promise of Justice

Promises may give purpose, meaning, and order to life, but they can also perpetually postpone the realization of hopes and desires. With *The Promise of Happiness,* Sara Ahmed suggests that desire is both what promises us something, what gives us energy, "and also what is lacking, even in the very moment of its apparent realization."[81] Ahmed demonstrates how the imposition of certain imagined futures—at the cost of others—can become an instrument of oppression. In other words, the imagination itself is a battlefield where unevenly distributed power shapes the fragmentation of shared dreamworlds.[82]

Some Indigenous theorists and activists are critical of the future-oriented temporality of justice in Western paradigms.[83] Potawatomi scholar Kyle Powys Whyte, for instance, notes a refusal on the part of Native American peoples to imagine "new" futures when climate injustice and its social impacts intensify existing imperial-capitalist regimes.[84] Symbolic and structural violence is embedded in some strands of international climate justice discourse, as influential institutions fail to address the uneven distribution of responsibility *for*, and vulnerability *to*, global warming and its deleterious effects.[85] The concept of justice itself has been critiqued by Anangax scholar Eve Tuck as an inherently "colonial temporality, always desired and deferred, and delimited by the timeframes of modern colonizing states as well as the self-historicizing, self-perpetuating futurities of their nations."[86] In West Papua, some Indigenous communities have given up on the future itself as an act of resistance—refusing the promissory hype of futures conjured by powerful forces.[87]

Within the realm of Western philosophy, Jacques Derrida is notable for celebrating justice as a universal and transhistorical force that will bring dramatic transformations to future horizons. Derrida describes elusive and lively specters of justice that contain "the attraction, invincible élan or affirmation of an unpredictable future-to-come (or even of a past-to-come-again)." In an attempt to protect this idea from the tools of deconstruction that he helped create, Derrida describes justice as something that is literally and figuratively empty—disconnected from "the *topos* of territory, native soil, city, body." Rejecting law (droit)—the application of existing rules—he suggests that justice requires us "to calculate with the incalculable."[88]

In reclaiming the promise of justice, and the idea of the future itself, it is important to refuse some of Derrida's more dramatic and seductive gestures. We find it necessary to ground dreams and struggles for justice in the topos of particular territories, soils, cities, landscapes, bodies, and technologies. We suggest: justice, like situated knowledge, is always *partial* in the sense of being for some worlds more than for others.[89] Rather than hold justice apart from situated political struggles, *The Promise of Multispecies Justice* identifies opportunities to deconstruct and reconstruct political positions, technical systems, ecological assemblages, and figures of hope.

As some resign the future to fate—by waiting for a definitive act of justice on future horizons—Indigenous leaders in North America are working toward improved relations in future lifetimes and generations (see Ishiyama and TallBear). Modest forms of justice are emerging through a process of co-becoming with biological and social relatives, even though not all of our relations are good relations. Indigenous thinkers in other parts of the world, like Benny Giay of West Papua, reclaim hope through "freedom dreams" that push against hegemonic forms of power as well as the limits of realism and realistic possibility.[90]

Counterhegemonic political imaginaries have long been guided by the articulation theory of Stuart Hall, who was originally interested in how different ideologies and institutions become joined together in contingent associations.[91] Articulation, in a general sense, means making speech sounds or linking things together. As a theory, Hall used the idea of articulation to understand how "people try to displace, rupture, or contest" dominant power structures. Counterhegemonic articulations involve getting inside ideological, institutional, and discursive formations to interrupt, transform, or change them from within.[92] Careful articulation work within the domain of justice has the potential to reinvigorate situated political struggles in the here-and-now, and also reorient promissory discourses toward worthy figures of hope in shared imaginative horizons.[93]

Some contributors in this collection are pushing the ideas of Stuart Hall beyond the realm of human discourse into the realm of ontology—to pursue justice through "the contingent, the non-necessary, connections between different practices" in multispecies worlds.[94] Jia Hui Lee brings us to urban Tanzania where an inventor is generating modest possibilities for justice by designing mechanical traps for pesky rodents. In the Philippines people are working toward forms of "molecular sovereignty"—or freedom from toxic chemical exposures—by setting up plants and animals as sentinel species. While reckoning with complex material entanglements, our contributors recognize that certain kinds of situated justice require an ethics of exclusion.[95]

Looking around, taking inventory of contemporary conditions, it is easy to slip into paralyzing feelings of despair. We live in an era of self-devouring growth, as Julie Livingston reminds us, with the expansion of capital markets undermining the conditions of life on Earth.[96] Atmospheric conditions are shifting, making it increasingly difficult for us all to breathe. As the privileged create cosmopolitan refuges in growing deserts, and safe bubbles in a viral pandemic, other peoples and creatures are living through endless wars. Industrial processes are uncoupling life from death, diminishing death's capacity to channel vitality back to the living.[97]

It is time to import the ecological principle of "intermediate disturbance" into dominant political institutions while creating the conditions for political assembly in alternative spaces.[98] Imagine a field of justice where multiple species circulate cradle to cradle—where the oikos of the household is in a dynamic equilibrium with interlocking ecological systems and economic circuits. Within this field, justice is slippery and spectral (see Govindrajan). Justice shifts and morphs across time, space, and species, resisting institutional capture or human mastery. It exerts an unpredictable force in the world as actuality or potentiality, through momentous events, everyday moments, and provisional judgments (see Marder; see Ihar).

A nomadic aesthetics of poaching has informed our curatorial and editorial practice. Taken together, the essays and poems in this collection tell a story that nimbly jumps scales and domains—moving from abstract speculation to situated political action and material intervention, and then back again. The authors show that it is possible to care for particular forms of life and biocultural communities, while at the same time holding onto promises of sweeping change on future horizons. Together we have developed an approach to multispecies justice that is anchored in the ongoing practice of being open and alive to the generative possibilities of each encounter.[99] This approach demands that we decide which dreams are worth dreaming—and by extension, which injustices are intolerable. It is also an invitation to renew our commitment to love, to live, and to fight for the possibility of flourishing in worlds present and yet to come.

Notes

1. Kirksey and Helmreich, "Emergence of Multspecies Ethnography," 546; Star, "Power, Technology, and the Phenomenology of Conventions," 43.
2. TallBear, "Why Interspecies Thinking Needs Indigenous Standpoints." See also Todd, "Indigenizing the Anthropocene."
3. Davis et al., "Anthropocene, Capitalocene, ... Plantationocene?"
4. Ursula Heise's 2015 keynote address on multispecies justice at the Green Citizenship symposium focused on "the political tensions between biodiversity conservation and environmental justice over the last thirty years, which have centered on the privileging of nonhuman species over the welfare of disenfranchised human communities." Ideas about multispecies justice were further developed by Subhankar Banerjee in 2018, "not as theory or analysis" but as "praxis." More recently, Danielle Celermajer and colleagues have examined how the entrance of other than humans onto the scene of justice challenges normative modes of liberal political discourse and practice. Meanwhile, T. J. Demos, Emily Eliza Scott, and Subhankar Banerjee's recently edited volume deploys a decolonial and interdisciplinary approach to widen considerations of justice as it intersects with contemporary art, visual culture, activism, and climate change. See Heise, "Multispecies Justice"; Banerjee, "Resisting the War"; Celermajer et al., "Justice Through a Multispecies Lens"; Celermajer et al., "Multispecies Justice"; Tschakert et al., "Multispecies Justice"; Demos, Scott, and Banerjee, *Contemporary Art, Visual Culture, and Climate Change*.
5. Dave, "What It Feels Like to Be Free."
6. Gay, *Book of Delights*.
7. Scarry, *On Beauty and Being Just*, 16.
8. Tsing, "Arts of Inclusion"; also see van Dooren, Kirksey, and Münster, "Multispecies Studies."
9. Rose and van Dooren, introduction to "Unloved Others"; also see Lorimer, "Nonhuman Charisma."
10. See, e.g., Kirksey, "Chemosociality in Multispecies Worlds"; Chao, *In the Shadow of the Palms*; Chao, "Can There Be Justice Here?"
11. Tsing et al., *Feral Atlas*. Wonder, writes Mary-Jane Rubenstein, is a disposition that encompasses an array of affective responses including awe, amazement, and marvel, but also dread, astonishment, and shock. Rubenstein, *Strange Wonder*, 9.

12 Greenberg, "When Is Justice Done?"
13 For more on these tactics, see Garcia and Lovink, "The ABC of Tactical Media."
14 See, for example, Bullard, *Confronting Environmental Racism*; Cole and Foster, *From the Ground Up*.
15 Taylor, *Toxic Communities*; Nixon, *Slow Violence*.
16 Allen, *Uneasy Alchemy*.
17 Shepard, "Advancing Environmental Justice."
18 Singer, *Animal Liberation*.
19 Kirksey, Schuetze, and Helmreich, "Tactics of Multispecies Ethnography," 3; Puig de la Bellacasa, "'Nothing Comes Without Its World,'" 208–9.
20 Appadurai, "Introduction," 17. See also Puig de la Bellacasa, "Nothing Comes Without Its World," 208–9.
21 Nussbaum, "Creating Capabilities"; Nussbaum, *Frontiers of Justice*.
22 Shrader-Frechette, *Environmental Justice*.
23 Eglash, "Introduction to Generative Justice."
24 Young, *Politics of Difference*.
25 Woolaston, "Ecological Vulnerability." On vulnerability as a framework for justice, see also Fineman, *Autonomy Myth*; Fineman, "Equality, Autonomy, and the Vulnerable Subject."
26 On egalitarian justice, see Locke, *Second Treatise of Government*; Anderson, "What Is the Point of Equality?"; Scheffler, *Equality and Tradition*.
27 On utilitarian justice, see Mill, *Utilitarianism*; Sidgwick, *Methods of Ethics*.
28 On contractarian justice, see Gauthier, *Morals by Agreement*; Rawls, *Theory of Justice*; Scanlon, *What We Owe to Each Other*.
29 Kim, *Dangerous Crossings*, 19–20.
30 Sousanis, *Unflattening*, 32. See also Gómez-Barris, *The Extractive Zone*, 11–12.
31 Ahmed, *Promise of Happiness*, 29.
32 Rutherford, *Laughing at the Leviathan*, 207.
33 Heise, *Imagining Extinction*, 199. See also de la Cadena, "Making the 'Complex We.'"
34 Haraway, *When Species Meet*, 244.
35 Fitz-Henry, "Distribution without Representation," 10.
36 De la Cadena, *Earth Beings*.
37 Stengers, *Cosmopolitics I*, 80.
38 Bishop, "Antagonism and Relational Aesthetics," 66; also see Laclau and Mouffe, "Hegemony and Socialist Strategy."
39 As the founders of the Science and Justice Research Center at the University of California, Santa Cruz, remind us, hegemonic and colonial efforts to build universalized knowledge and justice together can produce both epistemic and ontological forms of exclusion. Reardon, "On the Emergence of Science and Justice," 189–90. See also Reardon et al., "Science & Justice"; Stengers, *In Catastrophic Times*; Benjamin, "Black Afterlives Matter."
40 Clarke and Star, "Theory-Methods Package," 119.
41 Haraway, *Modest_Witness@Second_Millennium*, 75.
42 See, e.g., Weheliye, *Habeas Viscus*, 3. See also Bennett, *Being Property Once Myself*;

Jackson, *Becoming Human*; Wynter, "Unsettling the Coloniality of Being"; Benjamin, "Black Afterlives Matter."
43 Moore, Kosek, and Pandian, "The Cultural Politics of Race and Nature," 1–3.
44 See Roberts, *Fatal Invention*; Braverman, "Captive"; Copeland, *Cockroach*; Mavhunga, "Vermin Beings."
45 Nixon, *Slow Violence*; also see Muñoz, "Theorizing Queer Inhumanisms"; Puar, *The Right to Maim*; Martinez-Alier, "Environmentalism of the Poor."
46 Hernawan, "Torture in Papua."
47 Chao, "We Are (Not) Monkeys." See also Karma, *Seakan Kitorang Setengah Binatang*.
48 See, for example, Kafer, *Feminist, Queer, Crip*; Muñoz, "Theorizing Queer Inhumanisms"; Raffles, *Illustrated Insectopedia*.
49 Kim, *Dangerous Crossings*, 18. See also Anderson, "The Politics of Pests"; Glick, *Infrahumanisms*; Moore, Kosek, and Pandian, "The Cultural Politics of Race and Nature."
50 Boisseron, *Afro-Dog*, xiii–xx. See also Bennett, *Being Property Once Myself*.
51 Crenshaw, "Demarginalizing the Intersection of Race and Sex."
52 Tsing, Mathews, and Bubandt, "Patchy Anthropocene," S188.
53 See Peet and Watts, *Liberation Ecologies*.
54 Martinez-Alier, "Environmentalism of the Poor," 11–12.
55 Temper, Bene, and Martinez-Alier, "Mapping the Frontiers."
56 Crenshaw, "Demarginalizing the Intersection of Race and Sex."
57 Grossberg, "On Postmodernism and Articulation," 53.
58 Derrida, "Step of Hospitality/No Hospitality"; Serres, *Parasite*.
59 Stengers, *Cosmopolitics I*, 35.
60 Stengers, "Cosmopolitical Proposal," 999.
61 Laclau and Mouffe elaborated a theory of political subjectivity by building on Lacan, who suggested that subjectivity is not self-transparent and rational but irremediably decentered and incomplete. Laclau and Mouffe, *Hegemony and Socialist Strategy*, 125.
62 Bishop, "Antagonism and Relational Aesthetics," 66.
63 Bennett, *Being Property Once Myself*, 3.
64 Gómez-Barris, "Decolonial Futures." See also Chao, *In the Shadow of the Palms*.
65 Galvin, "Interspecies Relations and Agrarian Worlds," 244.
66 Davis et al., "Anthropocene, Capitalocene, … Plantationocene?," 5, 8.
67 Bennett, *Vibrant Matter*, 13 (italics added).
68 Povinelli, *Geontologies*, 55.
69 Kirksey and Helmreich, "Emergence of Multispecies Ethnography," 545.
70 Ingold, "Anthropology beyond Humanity."
71 Kirksey, "Species."
72 Dupré, *Disorder of Things*, 18.
73 Here we also build on Gregory Bateson's distinction between pleroma (the nonliving world that is undifferentiated by subjectivity) and creatura (the living world, subject to perceptual difference, distinction, and communication). Bateson, *Steps to an Ecology of Mind*, 489.

74 Shapiro and Kirksey, "Chemo-Ethnography," 482.
75 Sandra Harding has written about the possibilities of "strong objectivity" in the context of feminist standpoint epistemology. Kim Fortun's careful study of the Bhopal disaster at a Union Carbide pesticide plant in India stands as an exemplary account of molecular intra-actions and a social justice struggle that emerged from shared chemical exposures. Harding, "'Strong Objectivity'"; Fortun, *Advocacy after Bhopal*.
76 Reardon et al., "Science and Justice," 29. See also Reardon, "On the Emergence of Science and Justice"; Stengers, *In Catastrophic Times*; Benjamin, "Black Afterlives Matter."
77 Benjamin, "Black Afterlives Matter," 51.
78 Kimmerer, *Braiding Sweetgrass*; Whyte, "Our Ancestors' Dystopia Now."
79 Sousanis, *Unflattening*, 39.
80 Hurlbut, "Technologies of Imagination."
81 Ahmed, *Promise of Happiness*, 31.
82 Benjamin and Glaude, "Reimagining Science and Technology."
83 See, e.g., TallBear, "Failed Settler Kinship"; Winter, "Decolonising Dignity"; Hauʻofa, *We Are the Ocean*; Kimmerer, *Braiding Sweetgrass*; Stewart-Harawira, "Returning the Sacred."
84 Whyte, "Indigenous Science (Fiction) for the Anthropocene." See also Indigenous Action, "Rethinking the Apocalypse."
85 Tschakert et al., "Multispecies Justice"; Schlosberg and Collins, "From Environmental Justice to Climate Justice."
86 Tuck and Yang, "What Justice Wants," 6. See also Simpson, "Indigenous Resurgence," 21, 31.
87 Chao, *In the Shadow of the Palms*, 231–55; also see Indigenous Action, "Rethinking the Apocalypse"; Lear, *Radical Hope*.
88 Derrida, "Force of Law," 16; Derrida, *Specters of Marx*, 28; Derrida, "Marx and Sons," 253; Derrida, "For a Justice to Come."
89 Haraway, *Modest_Witness@Second_Millennium*, 37.
90 Benny Giay, a theologian and anthropologist from West Papua, has a long history of writing on hope and desire in Oceania. While earlier colonial tropes framed these hopes as "cargo cult" discourse, Giay has engaged with thinkers like Hannah Arendt and Walter Benjamin to insist that Indigenous peoples no longer "wait for outsiders to bring peace, happiness and justice." Quoted in Kirksey, *Freedom in Entangled Worlds*, 13. See also Kelley, *Freedom Dreams*.
91 Hall and Grossberg, "On Postmodernism and Articulation"; also see Clifford, "Indigenous Articulations."
92 Hall, "Signification, Representation, Ideology," 112.
93 Crapanzano, *Imaginative Horizons*; Kirksey, Shapiro, and Brodine, "Hope in Blasted Landscapes."
94 Hall, "On Postmodernism and Articulation," 53; Richland, "Jurisdiction."
95 Giraud, *What Comes after Entanglement?*, 2.
96 Livingston, *Self-Devouring Growth*.

97 On breathing, see Choy and Zee, "Condition—Suspension"; Hatch, "Two Meditations in Coronatime"; on cosmopolitan refuges, see Günel, *Spaceship in the Desert*; on vitality and death, see Rose, *Wild Dog Dreaming*.
98 See Dial and Roughgarden, "Theory of Marine Communities"; also see Garcia and Lovink, "The ABC of Tactical Media."
99 Cf. Barad, *Meeting the Universe Halfway*, x.

Glossary
Species of Justice

Sophie Chao and Eben Kirksey

Carceral justice involves imprisonment. This form of retributive justice entails punishing people for violating the law. Multispecies justice does not involve carceral justice but is instead allied with the prison abolition movement that envisions a world without cages or chains (see Lara).[1]

Climate justice foregrounds the disproportionately severe social, economic, health, and intergenerational impacts of climate change on vulnerable human groups. Advocates for climate justice suggest that functioning environments are a necessary condition for the fulfilment of other, intersectional justices—like environmental justice, social justice, and racial justice (see introduction).[2]

Competitive justice highlights the conflicts among different ideals about justice that proliferate within the political asymmetries, contestations, and forms of resistance that accompany the entanglement of species. By identifying conflicts between wildlife management and human flourishing, as well as forms of discrimination within the category of the human itself, identifying competition in the field of justice might help produce future coalitions within terrains marked by power differentials (see Paredes).[3]

Distributive justice is classically concerned with the distribution of benefits and burdens within society. Some approach this model with strict egalitarian principles, while others attend to how context governs the distribution of wealth and welfare. Emerging theories of bioproportionality expand the subjects of distributive justice beyond the human to effect an equitable partitioning of planetary resources across different species (see introduction).[4]

Ecological justice calls for recognition of other species as legitimate bearers of rights and recipients of resources. It seeks to develop institutional arrangements that can accommodate the claims and affordances of diverse creatures within its decision-making processes.[5]

Environmental justice aims to counter and redress the various forms of environmental discrimination that cause marginalized and racialized communities to bear the disproportionate burden of environmental harms, such as vulnerability to air pollution and water contamination and exposures to hazardous waste and toxic chemicals (see Ishiyama and TallBear; also see introduction).[6]

Generative justice is a bottom-up justice that emerges from collective resistance to hegemonic forces through peer-to-peer networks, open-source software movements, queer affiliations, agroecology projects, and Indigenous federations. This form of justice celebrates the consequential role of other life forms—bacteria, plants, fungi, and insects—within nested loops of multispecies regeneration (see Lee).[7]

Intergenerational justice is concerned with the scope and nature of relations, responsibilities, and obligations as these manifest across different generations.[8] It demands that institutions consider the potential impacts of their actions on the well-being of *future* human and other-than-human generations *and* their duties and responsibilities to generations past (see Lyons).

Multiworld justice is grounded in the phenomenology of matter itself. It approaches justice through the lens of lived experience, within and at the edges of dynamic worlds. In contrast to framing justice as potentiality, or what might be, multiworld justice attends to justice as actuality, or what just is (see Marder).

Participatory justice demands the equal, inclusive, and transparent participation of all parties in the development, enactment, and governance of justice-related institutions and practices. This framework addresses the challenge of reconciling disparate and often conflicting interests, values, and beliefs in achieving agreement over what counts as justice (see Lyons).[9]

Patchy justice materializes in fragments amidst uneven conditions of livability. Inspired by ideas about the "Patchy Anthropocene," this form of justice arises through grassroots oppositional movements, feral proliferations, and counter-hegemonic hopes that emerge in unstable places and uncertain times (see introduction; see afterword).[10]

Procedural justice involves due processes. Some believe that simply following the correct procedure results in an equitable outcome, regardless of whether distributive or restorative justice has ultimately been achieved (see introduction).[11]

Racial justice involves the fair treatment of people regardless of their racial identity. Beyond the mere absence of inequity and discrimination, racial justice calls for the establishment of institutional mechanisms that proactively sustain racial equity and reckon fully with the historical and colonial roots of racial violence (see Lara; see introduction).[12]

Recognition justice attends to how different beings gain or lose standing as a result of structural, institutional, cultural, legal, and economic regimes and attendant hierarchies of worth. It involves recognizing that past and ongoing legacies of unequal treatment, discrimination, and exclusion continue to produce just conditions of life for some and not others.[13]

Restorative justice is a response to injustice that focuses on restitution and resolution of issues arising from a crime or transgression. Mediation and conflict resolution can be used by victims, offenders, and a broader community to restore relations creatively (see Clark).[14]

Small justices are achieved through everyday incremental shifts and slight alterations rather than sweeping structural transformations. These micropolitical interventions, or microbiopolitical articulations, involve changes in daily patterns of thought and behavior.[15] Small justices operate in the middle of worlds, mediating rather than remediating multispecies relations. Often ambiguous and deceptively mundane, these little justices take difference and dissent as starting points for new cosmopolitical possibilities (see Ihar).

Social justice involves the fair and egalitarian treatment of all members of a given society in relation to questions of equity, access, well-being, participation, and rights. Movements for social justice are particularly concerned with achieving recognition, remedy, and redress for segments of society who are systematically marginalized (see introduction).[16]

Spectral justice troubles the boundaries between life, death, and the afterlife. The haunting force of ghosts can prompt the living to redress injury and repair damaged relations. Incomplete and inchoate in form and substance, spectral justice involves difficult negotiations across immanent and transcendent realms (see Govindrajan).

Substantive justice involves fair outcomes. Critical race theorists have considered how legal procedures often fail to account for substantive advantages and disadvantages at play across racial lines. Rather than focus on due process, or formal equality before the law, these theorists point to the substance of rulings when considering if justice has been done (see introduction).[17]

Transformative justice aims to achieve change in social, political, technical, or biological systems. Some victims of interpersonal violence or rape have used this approach to educate offenders and community members instead of pursuing punishment through the criminal or carceral system. Perpetrators of injustice may also seek to transform themselves in order to repair what they have damaged (see Clark).[18]

Transitional justice often entails fraught compromises as fragmented groups work to attain peace amidst large-scale armed conflict and human rights abuses. The process often involves recognizing, addressing, and remedying past wrongs through formal mechanisms including war-crime tribunals, truth commissions, criminal prosecutions, reparations programs, as well as efforts to support the physical and psychological healing of victims of war and violence (see Lyons).[19]

Notes

1 Brooks, "Retribution"; Hegel, *Philosophy of Right*. Compare with Bennett, *Being Property Once Myself*, 3–4, 8.
2 Schlosberg and Collins, "From Environmental Justice to Climate Justice"; Shue, *Climate Justice*.
3 Boisseron, *Afro-Dog*; Celermajer et al., "Justice Through a Multispecies Lens."

4 Rawls, *Theory of Justice*; Rawls, *Political Liberalism*; Mathews, "Bioproportionality."
5 Baxter, *Theory of Ecological Justice*; Baxter, *Ecologism*.
6 Pellow, *What Is Critical Environmental Justice?*; Schlosberg, *Defining Environmental Justice*.
7 Eglash, "Introduction to Generative Justice."
8 Thompson, *Intergenerational Justice*; Cooper and Palmer, *Just Environments*.
9 Gauthier, *Morals by Agreement*; Rawls, *Theory of Justice*.
10 Tsing, Mathews, and Bubandt, "Patchy Anthropocene."
11 Rawls, *Theory of Justice*; Nozick, *Anarchy, State, and Utopia*.
12 Richie, *Arrested Justice*; Ralph, *Torture Letters*; Garth and Reese, *Black Food Matters*; Shange, *Progressive Dystopia*.
13 Young, *Politics of Difference*.
14 Eglash, "Creative Restitution"; Heath-Thornton, "Restorative Justice."
15 Rousell, "Doing Little Justices."
16 Miller, *Social Justice*; Mill, *Utilitarianism*.
17 Matsuda et al., *Words that Wound*.
18 Morris, *Stories of Transformative Justice*. For an example of victim-led transformative justice, see Watson, "Silent Evidence."
19 Teitel, *Transitional Justice*.

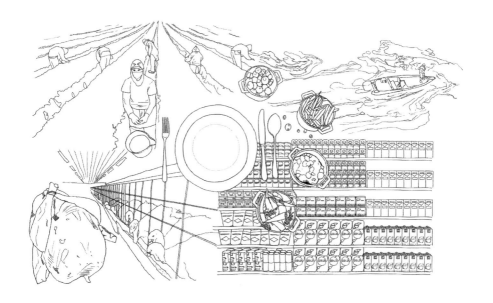

Blessing
Thanksgiving in the Plantationocene

Craig Santos Perez

FM.1 (*previous page*) Original drawing by Feifei Zhou.

Thank you, instant mashed potatoes, your bland taste
makes me feel like an average American. Thank you,

incarcerated Americans, for filling the labor shortage
and packing potatoes in Idaho. Thank you, canned cranberry

sauce, for your gelatinous curves. Thank you, native tribe
in Wisconsin, your lake is now polluted with phosphate

discharge from nearby cranberry bogs. Thank you, crisp
green beans, you are my excuse for eating dessert

à la mode later. Thank you, indigenous migrant workers,
for picking the beans in Mexico's farm belt, may your bodies

survive the season. Thank you, NAFTA, for making life so cheap.
Thank you, Butterball Turkey, for the word, "Butterball,"

which I repeat all day (say it with me): "Butterball, Butterball,
Butterball," because it helps me swallow the bones of genocide.

Thank you, dark meat for being so juicy (no offense, dry
and fragile white meat, you matter too). Thank you, ninety million

factory farmed turkeys, for giving your lives during the holidays.
Thank you, factory farm workers, for clipping turkey toes

and beaks so they don't scratch and peck each other
in overcrowded, dark sheds. Thank you, stunning tank,

for immobilizing most of the turkeys hanging
upside down by crippled legs. Thank you, stainless

steel knives. Thank you, scalding-hot de-feathering tank,
for finally killing the last still conscious turkeys.

Thank you, turkey tails, for feeding Pacific Islanders
all year round. Thank you, empire of slaughter,

for your fatty leftovers. Thank you, tryptophan,
for the promise of an afternoon nap.

Thank you, dear readers, for joining me
at the table of this poem. Please join hands,

bow your heads, and repeat after me:
"Let us bless the hands that harvest and butcher

our food, bless the hands that drive delivery trucks
and stock grocery shelves, bless the hands that cooked

and paid for this meal, bless the hands that bind
our hands and force feed our endless mouth.

May we forgive each other and be forgiven.

1

Spectral Justice

Radhika Govindrajan

1.1 (*previous page*) Original drawing by Feifei Zhou.

WE MUST HAVE MADE FOR AN ODD SIGHT, the three of us, gathered uncertainly around a corpse that lay on one side of a narrow mountain highway in the Kumaon region of the Central Himalayan state of Uttarakhand (see figure 1.2). The cars careening around the bend slowed down briefly to see if what we were surveying was of any interest before speeding away when they realized it was only a half-eaten animal carcass. One driver slowed down to offer a friendly piece of advice: "Don't hang around. A female leopard with two children lives here. They must have killed the bull."[1] Pushkar and I cast an anxious eye on the rocky hillside that loomed above us and agreed hurriedly that it was time to leave. We started to walk toward his car, urging his mother to join us, but she moved closer to the bull instead, covering her nose and mouth with her sari to hold back the unmistakable stench of decaying entrails. "Yes, it's definitely Gattu," she announced before asking me to come over and take some photographs.

Pushkar groaned when he heard this. "We're dead," he repeated to himself under his breath several times before getting into his car and shutting the door with a theatrical bang. I walked back to where Bina chachi (aunt) was squatting on the ground, as close as possible to the glossy black bull she was certain was actually Gattu. The bull's front leg stuck out at an unnatural angle, forced into the air by the swell of his distended belly brimming with microbial life. His face was partially concealed by the grass that had sprung up around the edge of the highway. A cloud of flies swirled around the exposed flesh in hypnotic whorls. An Egyptian vulture, with its distinctive yellow face and white ruff, sat patiently atop a nearby pine, waiting for us to leave so he could get a meal in before the leopards returned.

1.2 Gattu's corpse. Photography by Radhika Govindrajan.

Holding my breath, I crouched down by the steel guardrail to take photographs on my phone, focusing on the distinctive marks and features to which Bina chachi was pointing: the patch of auburn hair on his left leg where it met the hoof, the right horn bent at an odd angle, and the tuft of hair at the end of his tail. "He had a white mark on his chest," she repeated several times, "but the leopards have eaten it." I could feel the tremor in her hands as she reached out to anchor my own unsteady body. When we leaned over the guardrail to take a look at the bull's face, she sighed deeply. "Yes, it's definitely Gattu."

As we drove back to their village, Bina chachi tried to call Mahender, but he didn't pick up. "Why will he pick up now?" Pushkar asked tauntingly. "He's already fooled you good and proper." "Shut up, Pushkar," his mother snapped. "You and your father think everyone from my natal home is a thief." Mahender, she explained to me, lived in the next village from her natal family. She had run into him in the jungle that connected their villages three weeks ago, which was when he had offered a solution to a long-standing dilemma she had been struggling to resolve. Her cowshed had grown increasingly cramped after the recent birth of two calves. Gattu, her three-year-old bull, had not taken kindly to the new inhabitants of the shed and had injured one

Spectral Justice   **35**

of them. The aggression, chachi clarified, had intensified after Gattu started growing a hump and passed into adulthood. His personality, she added, had tended toward the bellicose (ladaku) right from the start.

Eventually, chachi had asked her sister-in-law if she could house Gattu in her shed until a solution was found, but this was only a temporary measure since the sister-in-law's cows would also eventually give birth. They didn't have enough money to build a new shed, and there were no local buyers to be found for a bull who was of no value except for breeding in this mountainous region where plough agriculture was uncommon. Further, the grueling labor involved in caring for these animals was starting to leave a physical imprint on chachi whose neck and back were permanently bowed from hauling loads of fodder up and down the mountainside.

Mahender had solicitously offered to facilitate a transaction between her and a man who lived close to his sister in the lowlands (where the terrain was suited to plough agriculture). The only catch was that she would need to cover the man's pickup rental and fuel costs given that he was, in a sense, doing her a favor by coming all the way to the mountains instead of buying a bull locally. On the face of it, Mahender's logic made little sense especially since the buyer would be getting Gattu for free already. However, the terms of the transaction were reflective of the situated history of rural livestock economies in Uttarakhand over the last two decades. State investment in rural dairy economies, particularly through the creation of artificial insemination centers throughout the mountains, had led to a remarkable increase in the population of dairy cattle. In mountain villages that were targeted by state officials for dairy schemes, many women, upon whose shoulders the bulk of the labor involved in raising animals fell, found themselves unable to cope with the additional burdens of care, specifically the "concrete work of maintenance" that accompanied the rise in the number of cattle in rural households.[2] The state's emphasis on dairy production as a key node of rural development thus not only created conditions for violence against cows, calves, and bulls but also intensified the social reproductive labor performed by rural women.[3] Chachi was only one of many women who navigated the dilemmas posed by increased care labor for livestock animals by reluctantly paying people to take the animals that they found themselves unable to care for. Those who could not afford to pay to have their animals taken away would abandon them in the forest, though not without sorrow or guilt.[4]

When the man from the plains came home to collect Gattu a week after her encounter with Mahender in the jungle, chachi was satisfied that he was taking Gattu to work in his fields. The man showed her pictures of the family's three cows on his phone. He told her they had a large shed and plenty of fodder and that Gattu would not lack for anything. Delighted, she paid the man fifteen-hundred rupees (approximately twenty-one dollars at the time) and the whole family—including her daughters-in-law, her sister-in-law, and her nieces—bid Gattu a tearful goodbye. Indeed, on the morning we came across the bull that chachi eventually identified as Gattu, she had asked Pushkar to stop the car near the carcass not because she had any suspicions that it was the bull she had raised from birth who lay eviscerated on the road, but because she, someone who loved cows with real fervor, wanted to bear witness to the bull's violent end, if only for a minute. It was only when she stood over the dead bull that she had the sinking feeling that the man who took Gattu might have been lying to her. Later that evening, once we were back at her home, she speculated that the man might not have been from the plains at all: "He must have been from some village in the mountains. Someone told me this is a business now. They take our bulls and our money, and then they abandon the bulls in the forest and pocket the money. This happened to Kusum Ram as well, do you remember? Her bull came back home after some days. But these bastards left Gattu far from home. He couldn't find his way back."

The "business" to which chachi was referring had become increasingly brisk in recent years as a result of the politics of cow protection pursued by Hindu supremacists who believe that cows must be protected from slaughter because they are the living embodiment of the deity Gau-Mata, the symbolic mother of all Hindus.[5] In the past, bulls and elderly or ailing cows might have ended up at a slaughterhouse in the lowlands, brought there either by brokers (like the fake buyer) or by butchers who were rumored to drive around the hills and pick up cattle abandoned by villagers who were no longer willing or able to care for them.[6] However, by the time we found Gattu in summer 2019 (and, indeed, even earlier), the ever-present threat of being lynched by Hindu supremacists meant that few people, especially Muslims and Dalits who were targets of vigilante violence, were willing to risk transporting cattle from the hills to the plains.[7] With nowhere to go, it became increasingly common to see abandoned cattle hanging out along state highways. Like Gattu, many of them were eventually killed and eaten by leopards.[8] This was an issue of growing concern not only for the state but also for ordinary Hindus like chachi who could no longer exist in a state of "ignorance" about their own role in the "journey" that bovines made from "cattle shed to dinner plate."[9]

Indeed, when we regrouped in the kitchen the next morning, chachi looked exhausted by the knowledge of the previous day. She told us she hadn't been able to sleep all night after being "frightened" (*darr*; a term widely used to signify what we might call haunted) by the sudden appearance of Gattu's glassy eyes in the middle of the night. "I should never have sent him with that man," she repeated several times. Her family tried to reason with her in different ways. While her sons were doubtful that she had correctly identified the half-eaten bull as Gattu (she rebutted their skepticism by comparing my photos of Gattu's corpse with photos I had taken of him during a visit six months earlier), her husband wanted her to file a police complaint against Mahender for cattle abandonment. Her daughter-in-law tried to placate her by saying that the men would be punished because *paap* (sinful actions) rebound against the person who commits it. However, chachi was inconsolable and continued to blame herself for having abandoned Gattu in her own "selfish" interest.

"Gattu's phitkar has definitely attached to me," she said, gloomily. "His shadow flitted around the room all night and troubled me." The Urdu term *phitkar* is usually translated as "curse," "malediction," or "censure."[10] When I asked chachi what she meant by it, she said, "Someone's curse [phitkar] attaches [to another person] when their spirit is unhappy." Gattu, she said, was clearly unhappy as evinced by his spectral presence in her room the night before. No matter who had abandoned him in the forest, it was she who was to blame for the bitterness that his untimely and violent death had caused him. "He is demanding answers from me," she declared. "That's why I've been frightened ever since I looked into his eyes."

When I asked if there was anything she could do to alleviate Gattu's phitkar, she said that she would need to get him nyaya, or justice. The concept of nyaya, Amartya Sen suggests, gestures to "social realizations" of justice that shapes the "lives people are actually able to lead."[11] Nyaya is what Aditya Malik, following Sen, understands as a "substantive," as opposed to merely "procedural," form of justice.[12] Bina chachi's vision of nyaya thus also entailed a form of justice that is situated in the social, and is deeply relational. Nyaya in this case demanded the work of repair: what Deborah Thomas calls an "obligation, not a gift," one that demands "active listening, mutual recognition, and acknowledgment of complicity."[13] For chachi, justice was substantiated through the repair of relationships between humans, animals, gods, and ghosts. She understood Gattu's haunting of her as an understandable outcome of her indefensible actions, which was why she had rejected her husband's suggestion to hand the matter over to the police in favor of interpersonally worked out nyaya.

The socially situated nyaya she sought could only be achieved through the intercession of the gods, what she described as devta nachana. Devta nachana, or making the gods dance, is a fundamental part of a Central Himalayan ritual practice called jagar, in which deities are beseeched, through singing and drumming by a group of ritual musicians, to occupy the bodies of their mediums.[14] The gods, once they are present among their devotees in embodied form, play a powerful role in resolving disputes, remedying curses, and delivering justice. Indeed, local deities such as Golu devta, who is colloquially known as nyaya devta or the god of justice, are hailed precisely for their ability to secure justice for unhappy and vengeful spirits or ghosts whose curse attaches to and troubles people.[15] It was to this divinely mediated form of seeking nyaya that chachi would turn as she attempted to come to terms with and atone for her complicity in Gattu's death and the "adversarial" turn their relationship had taken.[16]

This chapter takes Bina chachi's quest for nyaya for Gattu as the starting point for an exploration of the possibilities and limits of justice in circumstances where its absence or deferral is experienced as a prolonged haunting. Chachi's experience, I argue, invites us to consider the pursuit of justice as a spectral project. As a start, spectral justice is fueled by the grievances of ghosts who, to paraphrase Angie Morrill, Eve Tuck, and the Super Futures Haunt Qollective, refuse to be buried and insist on mattering.[17] Spectral justice is birthed from an accursed haunting that is, at its core, a claim to justice, a plea for the redress of an injury. Indeed, hauntings, as the anthropologist María Elena García so powerfully suggests, might be well understood as a spectral invitation to "repair" and "re-member" damaged relations with other-than-human kin.[18] They are, she asserts, acts of productive "disruption" that call on people to "reimagine an alternative present and future."[19] As Stuart Strange notes, the return of vengeful animal spirits serves the haunted a decisive reminder that "death creates inescapable obligations," extending kin relationships into the afterlife.[20]

Beyond its ghostly origins, however, what makes justice spectral is the fact that it is hauntingly incomplete and inchoate, its form and substance apparating and disapparating constantly as different ghostly and fleshy entities negotiate the terms of a just peace. Justice is spectral when it is haunted by the impossibility of its realization. It creeps in the shadowlands between the justice that ghosts want and the justice that humans are able to offer. It is "about something missing, about debt and haunting, about the insistence of a strange hollowness, palpable yet invisible."[21] Ultimately, understanding justice as spectral produces a reckoning with the difficult question of how one might live as "justly" as possible with ghosts while recognizing that a full and satisfactory justice might remain hauntingly elusive.[22]

In what follows, I trace the workings of spectral justice through chachi's attempts to assuage Gattu's unhappy spirit through rituals of repair. In particular, I examine debates between chachi and ritual practitioners who suggested that it was unlikely, if not impossible, for animals (particularly cattle) to take the form of unhappy spirits. Chachi responded to these practitioners and others who supported them, including members of her own family, by insisting that Gattu's spirit was, in fact, a powerful and persuasive otherworldly agent in search of nyaya. As far as chachi was concerned, Gattu was a "copresence"—an "active spiritual agent" whose "electric energy" was "felt, in, on, and around the body," to borrow the words of Aisha Beliso-de Jesús.[23] For her, Gattu was now part of the spirit world, alongside the many ghosts (bhoot), spirits (aatma), and deities (devta) who are ubiquitous and critical social actors in the Central Himalayan region.[24] She knew this with certainty, she maintained, because the embodied experience of Gattu's copresence was akin to that of other spirits by whom she had been troubled in the past.[25] This sensuous similarity allowed her to recognize that Gattu was expressing a demand for justice in much the same way as other spirits did: by troubling people with their uncanny and "seething presence" in the present.[26] She thus knew that the process of arriving at justice for Gattu would be as prolonged, fragile, and treacherous as it was when mollifying other ghosts and spirits.

One morning, a couple of days after we had stumbled upon Gattu, I came across chachi sitting on a slab of stone outside the bathroom in the lower courtyard. She was holding her little black and red cellphone to her mouth as she talked to her older sister. When I turned around to leave, she beckoned me over, shifting to make room for me on the slab. "You should talk too," she said, handing over the phone. After we had exchanged greetings and asked about each other's health, chachi's sister came directly to the point. "You need to go with her to Nanda Ballabh ji," she said. "He's a very powerful guru. An ascetic from Nepal comes upon [possesses] him."

When I hung up, chachi asked me to text her sister and ask for Nanda Ballabh's address. "If Pushkar doesn't have a booking [for his taxi] today, then let's go today itself," she announced. "I couldn't sleep all night." When I asked why, she told me that Gattu had troubled her all night: At about three in the morning, she awoke to the bed shaking violently. As soon as she opened her eyes, a tail lashed her face. She felt hot breath on her face. Two red eyes stared at her. As more came into focus, Gattu's head and half-eaten chest loomed above her. She closed her eyes in fear before feeling around for her phone. When the phone's light illuminated the room, the eyes were no longer there. She was paralyzed with terror and lay in a pool of sweat until she finally got out of bed at 4 a.m. to take her bath. "I kept turning around to look for the eyes," she said, with a shiver. "I'll have to do something. My sister says the guru will help. Don't let me forget to take rice."

People in Kumaon would often visit gurus who were mediums for powerful deities or babas (ascetics) to consult them on a wide range of problems, including unemployment, delayed marriage, land disputes, and illness, which were widely understood to have possibly been caused by haunting or curses.[27] On their visits, they would often take grains of rice from their own homes, the reasoning being that because they shared "both … [their] substance and … history with it, the … rice … [would] contain traces of that substance and that history."[28] The guru would toss these grains of rice up in the air, catch them in their hand, shift them from one palm to another, and then "read" and "interpret" the resulting arrangements on the basis of detailed conversations with their clients.[29]

Chachi was hopeful that consulting Nanda Ballabh would create a way to move forward and appease Gattu's undeniably unhappy spirit. The day after we found Gattu's body, she had finally managed to get in touch with Mahender, who spoke to the man in the plains and reported that Gattu had run away on the second day after arriving at his new home because the man's wife had not tied him tightly enough to the post. The man had sworn to Mahender that he had not abandoned the bull in the mountains and that he was still looking for him. Chachi could not tell if it was Mahender or the man or both who were lying to her, but did not know how she could arrive at the truth. Gattu's night visits revealed that her initial suspicions of treachery had been correct all along. Now, she needed to act quickly to relieve his curse.

Her urgency, in part, was motivated by her intimate knowledge of Gattu's bellicose and vengeful personality which, she was convinced, fueled his ghostly grievances. Of the many bulls she had raised, Gattu was the most quarrelsome, always fighting with cows, dogs, goats, and humans. He nursed grudges for days on end. On one occasion when he repeatedly headbutted and injured a neighbor, chachi hit him on the back in order to make him back off. Gattu, chachi recalled, waited four days to get his revenge. Her back had been turned when he stole up behind her and headbutted her into the terraced field below. "He would remember every small thing and take revenge," she said. "So why wouldn't he trouble me for something so big?" Indeed, as we got ready to leave, she told me that the ghost who visited her the previous night had exactly the same belligerent look that Gattu's face had always worn. It had been risky enough to turn one's back to Gattu in life. She certainly did not intend to ignore his ghost.

After gulping down breakfast, we set off on a two-hour journey to the guru's village. "You're wasting your money," Pushkar said, "Who in the world heard of a bull who came back as a ghost?" His mother was indignant. "Radhika, wait till Gattu's ghost gets after these fools. Just see how quickly they go running from deity to deity then." I hastily turned on the tape, hoping to avert conflict.

Fifteen minutes later, she fell asleep; I couldn't tell if it was from exhaustion or the anti-nausea medication she had taken right before getting in the car. "So, what do you think?" I asked Pushkar. "Will the guru be able to find a solution?" He was so focused on overtaking the truck that was farting gales of dark, acrid smoke at us that it took him a moment to answer. Now that he was not butting heads with his mother, he was more thoughtful in his response. "I don't know," he said, when we finally zoomed ahead. "Maybe she saw Gattu's ghost. But that happens in grief sometimes. She loves cows a little too much. She thinks, 'I am responsible.' That's all. There's no curse."

I was struck by how different Pushkar's understanding of this haunting was from his mother's. For chachi, it was Gattu's unhappy spirit who was the agent of her haunting. Gattu's "spectral speech" had established that he would not let her forget the ghastly circumstances of his death or the role that she had played in creating them.[30] But as far as Pushkar was concerned, chachi was haunted by her own grief, by her sense of complicity in Gattu's death. Really, she was haunted by an excess of love.

We finally reached the guru's village at noon. Pushkar elected to remain in the car, so it was just chachi and I who set off on a narrow path carved into the edge of the hill. The guru's house, a passerby told us, was nestled deep in the valley below. The morning's clouds had finally cleared up, making room for iridescent rays of sunshine that crept into every damp crevice. When we arrived at the house twenty minutes later, we joined a long line of folks who were awaiting an audience with the guru.

We soon struck up a conversation with an elderly couple in front of us. They told us that they had come from a nearby village to ask the guru for help with their grandson who had recently become an alcoholic and cut off all contact with his family after falling in with some gundas (goons). When chachi started describing her own predicament, they were immediately sympathetic. The grandmother exclaimed "Hai" with horror several times upon hearing about the red eyes and half-eaten chest, and she shook herself as if she too felt the press of Gattu's shadow. "I've heard of such things before," the grandfather said, resting on his stick. "When I was a child, there was a woman in the village who was caught by the spirit of some animal. It was very hard to expel. Animals demand justice too." Chachi concurred and said she hoped that the guru would find a solution.

After an hour, we were called into the guru's house. The room was dark, and it took our eyes a while to adjust after having stood in the bright courtyard for so long. "Tell me," the guru commanded as we sat down in front of him. After chachi recounted the events of the last few weeks, he put his hand out and asked for the rice that she had brought from home with her. He threw the grains up in the air, catching them easily despite the jerking of his body. After reading them in his palm, he began to speak of past family trouble with Masan, a malevolent deity associated with streams and water where bodies used to be cremated. "Yes, maharaja [king]," chachi replied. "I used to dance [when Masan came upon her], but we did a Masan puja [literally, ritual prayer; in this case, a complex ritual involving goat sacrifice] many years ago and we haven't had any trouble since." When the guru began asking about whether there were any women who had died unhappy deaths in past generations, she cut him off (much to his displeasure) and declared that their family had not been afflicted by a tiriya (unhappy, usually female, spirit) in years. "I've been afflicted by the bull's curse," she insisted.

"Don't interrupt," the guru shot back. "If you know it all, why come here?" Chachi was momentarily cowed, but then quickly recounted the events of this morning when she had awoken to find the bed moving and felt hot breath on her face before seeing Gattu's red eyes and half-eaten chest. "It's an unhappy woman's spirit," the guru declared. "A cow simply does not take the form of a ghost." Chachi pondered this for a minute but remained unconvinced. "It was him, maharaja. I know him." The guru did not respond but motioned us forward so that he could apply the ash he had just flicked from the burning logs onto a plate to our foreheads. After asking me a few questions about where I had come from and what I was researching, he turned his attention back to chachi. "Donate some money to a gaushala [cow shelter]," he said. "Prime Minister Modi ji himself does it... How could a cow ever trouble her children? She is a goddess... utterly pure. A false suspicion has taken hold of your mind. Take Gau-Mata's name, and this superstition will flee your mind."

The guru's words were in keeping with the emphasis on the maternal purity and largesse of the deity Gau-Mata (literally, Cow-Mother) in right-wing, cow-protectionist discourse.[31] I couldn't help but notice how he responded to chachi's specific question about her bull by speaking more generally about the divine power of cows at large (and Gau-Mata, more specifically). Perhaps this was because bulls pose a categorical dilemma to the politics of cow protection which hails the feminine, maternal innocence of the cow while remaining largely silent on the ill-treatment and slaughter of bulls.[32] Yet, while the guru might not have believed that bulls possessed the same divine power (shakti) as cows, he certainly did not believe that they could turn into ghosts either. Gattu's ghost, he insisted, was not real but a fiction of chachi's imagination, what he called superstition (andhvishwas). Interestingly, he did not question the existence of ghosts and spirits in general. Instead, what he refused to acknowledge was the possibility that cattle might traverse the boundary between the natural and the ghostly. Indeed, as I later found out, it was precisely this conservatism that made chachi question his authenticity as an oracle.

By the time we emerged from the room, the line of people waiting to meet the guru had stretched onto the narrow path. After squeezing our way past, we started the uphill climb to the car. The sun was no longer as welcome as it had been after the chill of the morning. Sweating, we walked in silence until we reached a point from where we could see the car. "I didn't think much of that guru," chachi announced. "A charlatan, I tell you. If it had been Masan or an unhappy woman's spirit, I would have known." I agreed that the guru had been dismissive of her concerns. "I must tell my sister what a liar he was," she said as we neared the car. "He clearly doesn't know anything."

Pushkar was asleep in the car when we opened the doors. "What happened?" he said groggily, adjusting his seat back to the upright position. When we told him about our conversation with the guru, he asked if he should stop at a cow shelter on the way back home. "Give them some money and put an end to this drama." His mother, already visibly irritated, snapped at the insensitive and ill-timed remark. "Cow shelter indeed! As if Gattu's ghost will disappear if I make a donation to some cow shelter. How is that justice for *him*? Tomorrow, I'll go and see Puran Ram in our village.

Radhika, Golu [the god of justice] comes upon him. He [Golu] will show us a path."

That evening, chachi was still fuming after her consultation with the guru. As she recounted the conversation to her daughter-in-law, Lata, it became clear she was most upset by the guru's claim that cattle could not take the form of ghosts or spirits because of their pure maternal form. "That man doesn't know anything," she repeated. "How could he say that?" she demanded. After all, she declared, any living being could take the form of a spirit if there was injustice involved in its death. "Even we acknowledge that the cow is a devi, but don't our devi-devtas [local gods and goddesses] take the form of spirits and ghosts when they're angered?"

Lata grunted her assent as she focused on kneading the dough for the evening's rotis. Chachi went on to tell us about the time that a popular local deity named Ganganath had afflicted her sister-in-law. It had happened a few years after her marriage, chachi recalled. Ganganath had come upon her sister-in-law in the form of a bhoot (ghost).[33] She would disappear in the forest for hours on end. The cows would come home from the jungle, but she would return only hours later. Some men said they had seen her sitting in trees. Ganganath had only let go of her when they had conducted an elaborate puja (complete with sacrifice) and built him a shrine.

Lata stopped kneading the dough to speak. "I talked to my brother today," she said. "I told him what happened with Gattu. He reminded me that our neighbor's cow had once been caught by a ghost. Strange voices would come from her. And her owners told us that her eyes would glow *red* in the dark of the shed. If a ghost can catch a cow, then why can't a cow become a ghost?"

"Yes. That's what I'm saying too. When a god can come upon a person in the form of a ghost, why can't a cow take the form of a ghost?" chachi demanded passionately. "Our gods and goddesses can make you dance if they want, they can drive you mad if they want. If cows are deities, then they can also become ghosts. Sometimes you have to become a ghost to demand nyaya."

Sometimes you have to become a ghost to demand justice. For chachi, establishing the fluidity of passage between ontological categories like cow, deity, and ghost was important because it was in the disordering and transcending of these categories that the possibilities of spectral justice resided. The fluid ontological landscape she sketched for us that evening was akin to what Sophie Chao calls "dispersed ontologies," social worlds in which "new and not-always-human actants... travel across and disrupt ontologies."[34] In this social world, humans and animals were "open and porous and vulnerable" not just to one another but also to gods, spirits, ghosts, and "other immaterial beings."[35]

What disturbed chachi about the guru's denial of these entanglements, specifically Gattu's ontological transformation from bull to ghost, was that it foreclosed the very possibility that Gattu was seeking spectral justice. In a landscape where spectral justice was so central to mediating relationships between a range of human and nonhuman entities, the guru's disbelief in a ghostly bull made it difficult for chachi to envision and enact what it would take to repair her ruptured relationship with Gattu. "Perhaps this didn't happen as much earlier," she mused much later that night, having made us (Lata and me) shift to her room because she was worried about Gattu troubling her if she fell asleep. "Perhaps that's why the guru didn't believe me. But then again, nobody abandoned their cows earlier. As the times keep changing, the ghosts keep changing."

"Don't do things in a rush." Bina chachi and I were sitting on a bed in Puran chacha's house. Opposite us, on another bed, sat Puran chacha (uncle) and his wife Asha chachi, both of whom were mediums for the gods Golu and Ganganath. The pair of mediums had never failed to secure restitution for the unhappy spirits and deities who had afflicted Bina chachi and other members of her family. She was now counting on their help in appeasing Gattu's angry spirit. She was, then, taken aback when Puran chacha advised her not to rush into a jagar. He told her that he knew she was feeling sad and that her sadness was understandable. However, and he asked her not to take his words the wrong way, it was possible that she had had a nightmare. "Sometimes bad incidents like this leave a mark," he said gently. "You feel scared, you feel sad."

Bina chachi was nonplussed. She struggled for words before saying that even she had suspected it was only a nightmare at first. But, on the second night, she has seen Gattu's eyes and half-eaten chest. She had felt the bed shake, felt his tail lash her face. She had *seen* and *felt* his ghost.

Puran chacha and Asha chachi rushed to assure her that they did not suspect her of lying. But it was, they remarked, important to have full knowledge of the matter before acting in haste.[36]

Bina chachi continued to insist that Gattu had come back as a ghost. Her certainty was rooted in past embodied experiences of haunting. She had first been caught by an unhappy female spirit in her natal home and then by Masan as a new bride. She knew, at an embodied level, what ghosts felt like. "Last night," she recounted, shaking with fear as she spoke, "I felt my bed shake. My face was so hot when I woke up. He was breathing over me. His shadow follows me all day. I can *feel* it … I *know* this is his ghost." This experience of embodied intimacy with Gattu's ghost had left her certain that his spirit would keep troubling her as long as it remained lost and wandering. She had done him an injustice, she said. Repair was necessary. A jagar would allow the gods to intercede, and divine intervention was the only way to ensure justice.

While Puran chacha and Asha chachi were sympathetic, they repeated their advice to take things slowly. They had never done a jagar for a ghost like this, they confessed. They needed to talk to people who might have experience with such matters. Choosing their words carefully, they added that time might also ease her affliction and obviate the need for a jagar. Either way, they concluded, it was best not to blunder in haste.

That night, as I reflected on the troubled and uncertain conversation that I had witnessed earlier, I found myself coming back to the question that Puran chacha had posed to Bina chachi on several occasions: *Why are you rushing things so much*? It was a reasonable question, I thought to myself. The process of undoing or healing affliction could often drag on for decades.[37] This meant that people were accustomed to living with ghosts and spirits over a long period of time. The fact that chachi had only felt the weight of Gattu's phitkar for a few days before visiting two gurus in as many days was unusual in a broader social context where the temporality of ritual action tended to be slow and prolonged.

Puran chacha's desire to slow things down also seemed related to the fact that he had not encountered an animal ghost before. Uncertainty about the ontological status of animal ghosts had led him to delicately ask if Gattu had appeared to chachi in a nightmare. Unlike Nanda Ballabh, he did not dismiss the possibility that Gattu had actually turned into a ghost; however, he was unsure if a jagar was the best path forward in order to establish the veracity of chachi's haunting by Gattu. In other words, Gattu's ghost had produced a serious categorical confusion that muddled and complicated chachi's desire to address an injustice.

In December 2019, six months after our initial visit to the medium's house, chachi told me that Puran chacha was still uncertain if an animal's phitkar could afflict humans. A further complication was that most of her family had refused to attend a jagar for what they thought was an elaborate (and expensive) drama fueled by guilt. She was upset by their lack of faith in her experience. "They tell me it's my own sadness [dukh] that's troubling me. But I know the difference between sadness and a ghost. I feel really bad about what happened. I admit that. But I *know* what a ghost is. I'm telling you, Gattu has come in the form of a ghost. It can happen. How can we know for sure that an animal cannot take the form of a ghost? Can't they feel sorrow? Can't they seek justice?"

I was struck by chachi's repeated emphasis on prior embodied experience as the basis of her certainty that Gattu had come back in the form of a ghost. In response to those who asked her how she knew that it was a ghost who afflicted her, she would declare that her certainty was based on past experience of having been haunted. To those who wondered if the ghost was Gattu or another unhappy spirit, she would say that she had raised Gattu from a baby and therefore knew that it was his quarrelsome and vengeful presence that stalked her. In insisting that it was Gattu—not her own guilt, not another spirit, but Gattu—who haunted her, chachi was, in effect, insisting that demands for spectral justice could originate from beings other than human and had to be taken seriously in order to live well with their ghosts. While this was proving to be more difficult than she had imagined, she was learning to live with Gattu's ghost while persevering in her efforts to hold a jagar so that she could hear his demands. "He still comes at night," chachi told me during a phone conversation in early 2020. When I asked if she found it hard to sleep, she said, "I do feel scared, but what to do? He won't go until he receives justice."

Notes

1. The conversations recounted in this article took place in Hindi and/or Kumaoni.
2. Puig de la Bellacasa, *Matters of Care*, 8. See also Gururani, "Forests of Pleasure and Pain"; Govindrajan, *Animal Intimacies*.
3. On the violence that dairy production inflicts on animals, see Cohen, "Animal Colonialism"; Narayanan, "Cow Is a Mother"; and Gillespie, *Cow with Ear Tag #1389*. On the social reproductive labor of women, see Vasavi and Kingfisher, "Poor Women as Economic Agents."
4. Media, animal-activist, and official discourse often represented these difficult decisions as driven by ruthless self-interest and a lack of genuine care for animals. However, as I have written elsewhere, such narratives did not capture the intense

guilt, doubt, and grief that rural dairy farmers experienced when letting go of animals for whom they could no longer shoulder the burden of care. See Govindrajan, *Animal Intimacies*; Govindrajan, "Labors of Love."

5 While it is beyond the scope of this essay to engage the cow-protection politics of the Hindu right wing, there is a rich body of scholarship on the topic. On the emergence of the cow-protection movement in the nineteenth century, and especially its centrality to an emergent Hindu nationalism that was dependent on "antagonism between Hindus and Muslims," see Van der Veer, *Religious Nationalism*, 3; Adcock, "Sacred Cows and Secular History"; Freitag, "Sacred Symbol as Mobilizing Ideology"; Robb, "Challenge of Gau-Mata"; and Yang, "Sacred Symbol and Sacred Space." For a feminist analysis of the gendered nature of the cow-protection movement as revealed through the maternal figure of Gau-Mata, see Gupta, "Icon of Mother." For a detailed exposition of the postcolonial history of cow protection and its relationship to both the secular state and the Hindu right wing, see Chigateri, "Negotiating the Sacred Cow"; Copland, "Cows, Congress, and the Constitution"; De, "Cows and Constitutionalism"; and Jaffrelot, *Hindu Nationalist Movement*. For an account of the effects of cow protection on Dalit, Muslim, and Adivasi communities in India and their response to this violence, see Ghosh, "*Chor*, Police and Cattle"; Gundimeda, "Democratisation of the Public Sphere"; Dhar and Jodhka, "Cow, Caste and Communal Politics"; Imran, "Impact of 'Cow Politics'"; Kapoor, "'Your Mother, You Bury Her'"; and Tayob, "Disgust as Embodied Critique."

6 The government of Uttarakhand had passed a law criminalizing cow slaughter, the consumption of beef, and the transportation of cattle across state lines as early as 2007. However, much to the despair of animal-rights activists and gau-rakshaks (cow-protectors) who patrolled state highways and borders alongside the police, the "smuggling" and slaughter of cattle remained largely unabated in the years following the passage of the law (Govindrajan, *Animal Intimacies*). The victory of the Hindu right-wing Bharatiya Janata Party (BJP) in the 2014 national election played an important role in changing this situation. The BJP had made the protection of cows from slaughter a major electoral issue. Indeed, one of Narendra Modi's promises was that he would put an end to the "pink revolution" (that is, the slaughter of cows, pink being the color of their flesh) if elected prime minister. Vigilante violence against Muslims, Dalits, and Adivasis who were suspected of smuggling or slaughtering cows surged in the years after 2014; vigilante cow protectors came to form a parastate, aided and abetted by state authorities (Jaffrelot, "Hindu Nationalism"). This vigilante violence was accompanied by a stream of legislation to control the trade in cattle and prevent slaughter. It was in the context of the growing power of right-wing Hindu politics that the trade in cattle in Uttarakhand shrank dramatically in the years following 2014 (Govindrajan, *Animal Intimacies*).

7 In 2017, the website India Spend noted that "Muslims were the target of 52% of violence centred on bovine issues over nearly eight years (2010 to 2017) and comprised 84% of 25 Indians killed in 60 incidents" of cow-related violence. "As many as 97% of these attacks were reported after Prime Minister Narendra Modi's

government came to power in May 2014." Abraham and Rao, "84% Dead in India's Cow-Related Violence."

8   People only half-jokingly noted that the growing number of abandoned cattle meant that leopards no longer had to hunt wild animals to feed themselves. Videos of leopards killing cows on the highway at night were widely circulated in Whatsapp groups fueling jokes that leopards were the only ones who could still eat beef in India. In 2020, when news circulated that a minister in the Legislative Assembly of the state of Goa had called for big cats who killed cattle to be punished, people in the mountains joked that all their leopards would be in jail if this ever came to pass.
9   Staples, "Blurring Bovine Boundaries," 1128.
10 Wagenaar et al., *Transliterated Hindi-Hindi-English Dictionary*, 483.
11 Sen, *Idea of Justice*, 82, xv.
12 Malik, "Darbar of Goludev."
13 Thomas, "Rights, Gifts, Repair," 6.
14 For more on the jagar, see Fiol, "Dual Framing"; Jassal, "Divine Politicking"; Leavitt, "Meaning and Feeling"; and Sax, *God of Justice*.
15 The deity Golu is worshipped across castes and communities in Kumaon. The history of his emergence as a deity emphasizes his inherent fairness and consequent ability to dispense justice. Golu's devotees beseech him to intervene in disputes in one of two ways: either through the embodied rituals known as jagars or through petitions (*manautis*), which are hung in his temple and written in a "hybrid legal and devotional language." Malik, "Darbar of Goludev."
16 Buyandelger, "Asocial Memories."
17 Morrill, Tuck, and the Super Futures Haunt Qollective, "Before Disposession," 1, 7.
18 García, "Landscapes of Death," 17.
19 García, "Landscapes of Death," 18; see also García, *Gastropolitics and the Specters of Race*.
20 Strange, "Vengeful Animals," 140.
21 Klima, *Ethnography #9*, 9.
22 Derrida, *Specters of Marx*, xvii–xix. See also Wijaya, "To Learn to Live with Spectral Justice"; García, *Gastropolitics and the Specters of Race*.
23 Beliso-de Jesús, *Electric Santería*, 3. See also Gold, *Spirit Possession*; Ong, *Spirits of Resistance*.
24 On the role of gods, spirits, and ghosts in the Central Himalayas, see Govindrajan, "Adulterous Dotiyal"; Klenk, "Seeing Ghosts"; Jassal, "Divine Politicking"; and Sax, *God of Justice*.
25 I use the word *embodied* to gesture to the fact that spirits and ghosts often share bodies with those whom they haunt.
26 Gordon, *Ghostly Matters*, 8.
27 See Jassal, "Divine Politicking"; Sax, *God of Justice*.
28 Sax, *God of Justice*, 56.
29 Sax, *God of Justice*, 56.

30 Kumar, "Strange Homeliness."
31 Historians date the emergence of the mother-goddess Gau-Mata to the cow-protection movement of the late nineteenth century. Her iconic representation is as a white, bejeweled cow whose body is studded with Hindu deities. As the historian Charu Gupta notes, religious nationalists in the late nineteenth century argued that the killing of a cow was "matricide" because the cow was a mother who nourished her (Hindu) sons and, by extension, a Hindu nation with her milk: "The material body of the mother cow was equated with the Hindu nation, where she was the benevolent mother, whose womb could provide a 'home' to all." "The Icon of Mother," 4296.
32 Indeed, as Narayanan notes, the logical outcome of the Hindu right wing's celebration of cow-milk as the substance of kinship between Hindus and Gau-Mata (and their subsequent emphasis on dairy production) is the butchery of unproductive male bovines. "Cow Is a Mother."
33 Ganganath is a local deity who is known to be capricious and easily angered. He often afflicts women in the form of a ghost.
34 Chao, "In the Shadow of the Palm," 634.
35 Fernando, "Supernatureculture." See also Bubandt, "Anthropocene Uncanny."
36 As Sax notes, gurus place a great deal of emphasis on having full information from their clients. Such consultations offer involved detailed questioning to understand the exact nature of the haunting. *God of Justice*, 56.
37 Govindrajan, "Adulterous Dotiyal"; Sax, *God of Justice*.

2

Rights of the Amazon in Cosmopolitical Worlds

Kristina Lyons

2.1 (*previous page*) Original drawing by Feifei Zhou.

Who Cares for a "Common Home"?

THE "FIRST REGIONAL FORUM on the Rights of the Amazon: Our Common Home" took place on November 22–23, 2018, at the Experimental Amazonian Center on the outskirts of Mocoa, Putumayo. The event was held in a kiosk inspired by ancestral maloka (Indigenous meeting house) architecture and constructed in the middle of an artificial lake. Housed in pens, fenced-in areas, and tanks around us were a jaguar, tigrillo, eyra cat, wild pigs, toucans, anacondas, monkeys, and an array of aquatic life that had been rescued from illegal traffickers and were in recovery for subsequent return to their habitats in the selva. On the stage, Indigenous, campesino, and Afro-descendent social leaders were assembled to share their perspectives on Sentence 4360. This landmark sentence, dictated by the Supreme Court of Justice in Bogotá, Colombia, in April of that same year, recognized the country's Amazon Basin—which comprises around 35 percent of the national territory—as a subject of rights entitled to legal protection.

The event was organized by Corpoamazonia, one of the three environmental authorities with jurisdiction over the Colombian Amazon. Representatives of the region's rural communities had been invited to speak after several institutional presentations explained the structural obstacles that Corpoamazonia faced in implementing the orders dictated by the court in its ruling. When an Indigenous elder from Leticia, Amazonas, interjected, he blatantly disclosed that he did not know why he had been invited to the event. He sat with his arms crossed over his chest, seemingly unimpressed by the commotion created around the legal ruling. Until that day, he had never heard of the sentence, much less its legally binding obligations for the region's rural communities. His comments lay bare the crux of the exclusions reproduced by a court sentence that was dictated from the Andean capital of Bogotá with minimal participation of local communities or regional actors.

These complications extend far beyond a mere lack of dissemination of information. They perpetuate the drawn-out structural violence of an Andean-centric country that continues to treat the Amazon as one of its territorios nacionales (national territories) historically administered under tutelage by a special office in the central government. Margarita Serje writes that these "savage territories"—sparsely populated, generally Indigenous, and far from the capital and other urban centers—were first converted into missionary outposts, then agricultural frontiers and fronts of colonization chronically problematic for the state.[1] Later they became known as "zones of public order," a euphemism for violent epicenters of over five decades of social and armed conflict. Historically, the region's inhabitants have been treated reductively as objects of intervention or perpetual problems to be solved—not as protagonists of their realities with valuable know-how, transformative potentiality, and constitutional rights to participate in territorial planning and environmental governance decisions.[2]

Corpoamazonia is widely distrusted. Small-scale cattle ranchers throughout Puerto Guzmán, Putumayo, spoke of being bombarded by pamphlets dispersed from military helicopters threatening to arrest anyone caught felling trees to expand pasture. They were convinced that Corpoamazonia dropped jaguars with GPS tracking devices near their farms to monitor and arrest them when they shot the animals after they devoured their cattle. It was not lost on them that their ranching had produced the loss of forest habitat, and hence, of the felines' food sources. They felt caught in a vicious cycle. They were receiving no subsidies for forest conservation or technical assistance to convert to silvipastoral practices. Meanwhile, the state continued to sign off on more concessions to multinational oil companies operating in the Amazon.[3]

A Corpoamazonia official emphasized that the environmental authority would need to rely on community-based monitoring and control to combat deforestation. He spoke about the need to fund local environmental committees within existing campesino community action councils. He said this right before a representative of the industrial mining sector gave a presentation about the impossibility for municipalities to deny access to subsoil mineral wealth. A firm "no" to mining was out of the question, he explained. The contradictory messages of the forum were disconcerting. By the end of the event, I found myself increasingly preoccupied by the paradoxes and risks produced by a legal case that was not driven by the initiatives of local actors. Court orders about the "rights of nature," dictated from the country's Andean capital, obligated already marginalized communities to assume new responsibilities even as environmentally destructive industries intensified activities in the region.

The legal sentence that recognized the Colombian Amazon as a subject of rights produced excesses and aspirational horizons. By excess, I think with Marisol de la Cadena's concept to consider "the limits of" progressive visions in general, and of rights-of-nature paradigms in nascent climate jurisprudence in particular.[4] Given that excesses tend to be ignored, de la Cadena argues that ignorance—both as disavowal and as incapacity to know—cancels that which is neglected, rendering it invisible or nonexistent.[5] Attempts to expand legal fields through the recognition of new subjects and more capacious rights frameworks are not immune to perpetuating this form of ontological violence when they deny the voices and realities of communities most implicated in the impacts of their judicial decisions.

Conventionally, public policies designed in Bogotá have attempted to eradicate conflicts in the Amazon through a combination of so-called law and order (i.e., the violence of security). Considering this rights-of-nature case in the presence of regional inhabitants of the Amazon, I suggest, might generate a bettering of socioenvironmental conflicts instead of an aspiration to pacify them violently, on the one hand, or manage them administratively, on the other. For feminist philosopher Isabelle Stengers, "bettering" conflict involves learning to "slow down the cunning of reason."[6] This slowing down entails questioning the assumption that one is imbued with the authority to know the problem that needs resolving. It also implies slowing down impulses to pacify frictions or resolve incommensurable differences through moralizing acts of tolerance that end up flattening these differences and, consequently, ignoring a series of actors and realities involved in the situations under dispute. How does one become attentive to actors whose existence has been systematically denied? How can possibilities for dialogue be constructed without a reliance on being "in common agreement"?

Rather than imbuing the law with the prescriptive mastery to resolve socioenvironmental conflicts, something like justice can be constructed through a process of dialogical, cosmopolitical copresence. This does not mean replicating conventional spaces of dialogue that tend to be organized as informational sessions after technical decisions have been made or monitoring processes that imply some level of formal participation at the implementation stage of a project or policy. Nor does it mean enforcing prior consultation mechanisms when these are used to deny the possibility to oppose or drastically reorient decisions and terms of debate. Instead, I call for processes of continuous interethnic dialogue and informed citizen-led participation in all stages of public policy making, territorial planning, and environmental governance. Multiple worlds are at stake in what is singularly referred to as the Amazon Basin. Cosmopolitical justice might be achieved, at least provisionally, by engaging with more territorial actors and local realities (human and other-than-human—jaguars, remerging secondary forests, cattle, coca plants, sacred sites, and ancestral spirits, to name only a few of the beings and socioecological relations that form part of the territories of the Basin).

The ideal of justice remains elusive in Colombia. Transitional justice activities are resulting in criminal prosecutions, truth commissions, and reparations programs. Conflicts persist, especially in the country's frontier regions—even after the signing of the peace accords in 2016 between the national government and the once largest guerrilla organization in the Western hemisphere, the Revolutionary Armed Forces of Colombia-People's Army (FARC-EP). The ongoingness of transitional justice as it overlaps with biocentric legislation provokes a series of questions: When can we say that justice has materialized, and for whom has it been achieved? Which understandings of justice are we referring to when we seek to repair socioecological relations and the protection of ecosystems that have suffered sedimented layers of damage? Have decades of war and new vulnerabilities after the post–peace-accord process foreclosed the possibility of anything beyond a precarious peace? What are the variations of justice in each particular place and among diverse rural and urban communities that share territories and histories of violence?

In this chapter, I explore these questions in conversation with other queries about the substance, implementation, and extension of rights of nature. I reflect on what happened to court orders dictated for the rights of nature of the Amazon once the courts technically abandoned the case, as well as the impacts of their decisions on state entities, civil society, regional social movements, and public opinion. In particular, I interrogate the roles imagined for local communities in the transformation of territorial relations and the implementation of climate change mitigation strategies. I then discuss the new risks and forms of criminalization that may emerge when environmental conservation mechanisms change (or do not change) hands from illegal armed actors to police and military forces. I conclude by offering propositional reflections for bettering conflict. This begins with cosmopolitical dialogues that attend to the complex realities influencing the dynamics of deforestation and environmental degradation in the Colombian Amazon.

Proliferating New Rights for "Nature"

"A judge today has to be avant-garde," Judge Tolosa tells me in his marked santandereano accent. "Judges normally magistrate over the past, not the future. The law follows behind social decisions. How can the power of the decisions made by judges be maintained after the construction of new rights? They need to have teeth to be materialized." We are seated in Tolosa's chambers in the Supreme Court of Justice in Bogotá in May 2019, a little over a year after he dictated the historic sentence that was later promulgated by Corpoamazonia. The judge toys with his glasses momentarily. He is a thin man with a full head of black hair that makes him appear younger than most other Supreme Court magistrates. "Environmental problems require technical evidence. They cannot remain in the abstract," he says, referring to the links established between climate change and constitutional rights to human health. Judge Tolosa weighed all of this in the findings of the sentence.

Sentence 4360 was dictated on April 5, 2018, in response to a case led by the Bogotá-based legal "think-do-tank," Dejusticia, the Center for Law, Justice and Society. Dejusticia enlisted twenty-five children and youth from around the country, aged seven to twenty-five, in a lawsuit against the president, the Ministry of Environment and Sustainable Development, the Ministry of Agriculture and Rural Development, and the forty municipalities of the Colombian Amazon. The lawsuit claimed that their rights to a healthy environment, life, food, water, and health were threatened by the government's failure to control deforestation. Despite Colombia's national and international obligations and voluntary commitments made in climate summits, the country was contributing to climate change. The plaintiffs argued that the government was obligated to reduce deforestation through at least three commitments: (1) the Paris Agreement committed Colombia to reduce greenhouse gas emissions; (2) a joint statement of Colombia, Germany, Norway, and the United Kingdom bound the government to reach zero net deforestation in the Amazon by 2020; and (3) domestic law 1753 (passed in 2015) required the government to reduce the national annual deforestation rate.

Judge Tolosa examined the 1991 constitution and found that its concepts, along with previous jurisprudence, international law, and academic scholarship, elevated a healthy environment to a fundamental right. His court confirmed that the government had not dealt effectively with deforestation and climate change despite its obligations. This analysis was based on the principle of solidarity in Article 1 of the Constitution, which refers to the solidarity between people. In the case of the Amazon, the court argued that it was necessary to consider "the other" in this principle of solidarity—that is, others who also inhabit the planet, such as plants, animals, and future generations that deserve to enjoy the same environmental conditions as today's generations. The precautionary principle further empowered the court to limit possible actions of present generations through an obligation to "do no harm" and instead to care for and guard over natural resources and the future human world.[7]

The court then ordered the government to develop a series of action plans within five months of its decision. These included an Intergenerational Pact for the Life of the Colombian Amazon (PIVAC) and strategies to reduce deforestation to net zero, combat greenhouse gas emissions, and update municipal territorial ordinance plans throughout the Basin. By declaring the country's Amazon a subject of rights, the court sought to advance the emerging field of biocultural rights.[8] These rights are intrinsically linked to the ancestral practices of local communities, their territorial roots, perceptions of their lifeworlds, and respect for reciprocal relations with nature that provide for their daily sustenance. Inspiration came from the philosophical constitutional principles outlined in the first rights-of-nature case in Colombia (Sentencia STC 622) when the domestic Constitutional Court granted the Atrato River legal personhood in November 2016.[9]

The idea that nature possesses inalienable rights akin to human rights has evolved from a strictly theoretical concept to the basis of policy changes in several countries and US municipalities in the last two decades. Ecuador grabbed international headlines when it included a rights-of-nature framework in its 2008 constitution. These celebrated Ecuadorian legislative provisions drew on earlier work by the Community Environmental Legal Defense Fund that supported citizens of Tamaqua, Pennsylvania, to formulate the world's first local rights-of-nature ordinance in 2006. Scholars have written about rights of nature for decades, but global networks to develop and promote Earth Jurisprudence, or a philosophy and practice of law that recognizes the inherent rights and value of nature and the interconnectedness of the planet's living systems, only emerged in the 1980s, propelled by Indigenous movements in the Global South. Throughout the early 2000s, institutions and centers for Earth Jurisprudence formed in the UK, South Africa, Australia, New Zealand, the United States, and elsewhere.[10] Identifying the limits of environmental law, which traditionally has been a regime of permits and licenses to regulate the management of natural resources, rights of nature and Earth Jurisprudence frameworks claim to focus on the diverse relations that communities sustain with/in their territories—relationships they argue the law was previously unable to recognize.[11]

In Colombia, a series of rights-of-nature cases have proliferated since the 2016 Atrato River case. A number of river basins—La Plata, Coello, Combeima, Cócora, Cauca, Magdalena, Pance, Otún, and Quindío—were recognized as entities subject to rights for protection, conservation, maintenance, and restoration by the state. A 2017 case recognized the emblematic oso de anteojos (speckled Andean bear) as a subject of rights, expanding previous animal protection legislation that classified animals as sentient beings and guaranteeing their protection as a part of biodiversity. A year later, a tribunal court declared the páramo, or alpine tundra, of Pisba to be a subject of rights after national policies ordered the delimitation of these ecosystems to ensure their protection from industrial mining and other economic activities.[12] Most recently, the courts declared the Isla de Salamanca Parkway and Los Nevados National Park subjects of rights, citing the state's failure to protect these strategic ecosystems from industrial activity, agriculture, deforestation, and overall environmental degradation.

This recent biocentric legal turn builds on environmental protections achieved in Colombia's 1991 constitution and the long-standing struggles of Indigenous, Afro-descendent, and diverse rural communities in defense of their territories and the construction of consulta previa (prior consultation) rights.[13] They also respond to growing global and national concerns over the anthropogenic force of climate change, mass deforestation, and species loss amid the country's ongoing transitional justice process. Across different national contexts, issues of war crimes and violence have primarily focused on humans as victims and actors within human rights and humanitarian frameworks. In Colombia, however, there is growing public and legal debate over how soils, rivers, forests, and territories are also casualties of war, requiring criminal prosecution and reparative treatment in the transitional justice scenario.[14] In 2019, the Investigation and Accusation Unit of the Special Jurisdiction for Peace, the tribunal established as part of the transitional justice mechanisms, recognized nature as a "silent victim" of the country's armed conflict, taking inspiration from rights-of-nature sentences and the worldviews of Indigenous, Afro-descendent, and campesino communities.[15]

Judges of the End of the World

Over a cup of coffee and aromática in his chambers, Judge Tolosa enthusiastically leads me through the innovative legal aspects of the sentence he dictated for the Amazon. He comments on its contribution to climate change jurisprudence. The ruling recognizes the rights of future generations, particularly their right to be heard in the formation of policies that will affect them. It also advances discussions about the rights of nature and helps concretize interpretations of the Paris Accords. Midconversation, Tolosa references an opinion piece published in Colombia's *Semana* magazine, provocatively entitled "Jueces del fin del mundo" (Judges of the End of the World). The essay characterizes the future-oriented stakes and planetary scales at play in emerging climate jurisprudence.[16] Despite its largely hopeful tone, the article cautiously underlines the perverse logic of such jurisprudence. A legal system, inherited from colonialism, has permitted large-scale plundering and environmental degradation in the name of capitalist growth. Now, the courts are attempting to hold actors accountable for curtailing the phenomena that they produced. Weak institutional capacity to implement existing laws and newly established rights, alongside pervasive political corruption, compound more typical questions that have emerged in global rights-of-nature debates.

Ontological claims are being made about legally recognizing living beings that were previously regarded as a bundle of ecosystem services. Potential litigants are working through issues of proxy (who will speak on behalf of nature) now that new entities have been granted rights. Many wonder whether criminal and public law provide more robust protection than rights-based moves. Newly conceived ideas of legal personhood are vulnerable to swift challenge or reversal, particularly given that corporations also have legal standing as persons. Critics have asked: Is it desirable that the Niger Delta might exist in legal affinity with Shell? Where does a forest materially and symbolically begin and end, in terms of its "being"?[17] As we discuss the complex historical, political-economic, and socioecological realities of the country's Amazon, Judge Tolosa reluctantly acknowledges, "There is a problem in the construction of the sentence. It has the local problematics hidden within it."

After our conversation, I was left pondering a series of paradoxes. Not least was the question of what is ignored or denied when it is claimed that magistrates are now making judgments in an era of civilizational crisis that may bring about the "end of the world." I reflected on a recent visit I made to a Nasa Indigenous resguardo (reservation) on the outskirts of the municipal seat of Puerto Guzmán, Putumayo. The Nasa elders spoke of their arrival to Amazonian territory from their ancestral lands in the Andes due to land concentration and dispossession provoked by the war. Now, they mostly live on open pasture and they explained how the state provides no systematic technical assistance to rural communities based on Amazonian agroforestry systems. Their experience was not unlike that of the many waves of campesinos who settled in the region since the 1930s, a product of military colonization propelled by the Colombia-Peru war (1932–33), expulsion from the country's interior during years of bipartisan violence (1948–58), or motivated by the spur of a protracted history of extractive boom-and-bust economies, the last of which pivot around illicit coca crops, crude oil extraction, and illegal gold mining.[18]

Indigenous peoples, of course, including Andean-Amazonian Native peoples, understand the present time as dystopic. They approach climate change as having already occurred through the irreparable transformations of their territories and lifeworlds induced by colonial violence.[19] Furthermore, the seemingly new constitutional and epistemological constructs that recognize rights of nature have emerged alongside much older phenomena. These biocentric legal moves take orientation from ancestral practices, political structures, and the jurisprudences of Indigenous peoples that persist despite the hegemonic structures of liberal constitutionalism and ongoing colonial practices.[20] The lack of recognition of the contribution of these jurisprudences and life philosophies in the building and implementation of rights-of-nature cases is one of the core problems of Sentence 4360.

Moreover, the country's Andean-Amazonian foothills and plains sit at an intricate political, economic, and legal crossroads. As a historic epicenter of internal social and armed conflict, the western part of the Amazon is impacted by the complex geopolitical nexus generated by illicit crop production, narcotrafficking, and the US-Colombia war on drugs. In contrast to the country's eastern Amazon, which is more extensive in territory and less densely inhabited with a largely dispersed Indigenous population, the western region was converted into a settler zone for displaced and marginalized peoples from the country's interior. These more contemporary waves of violent territorial reconfiguration were layered on colonial dispossession of ancestral Indigenous lands. All of these dynamics have contributed to impoverishing rural communities, fomenting illicit activities, and creating conditions for enduring territorial disputes between armed actors.[21]

The municipality of Puerto Guzmán, for example, has experienced intense deforestation following the peace accord between the government and

2.2 Deforestation in Puerto Guzmán, Putumayo.
Photograph by Jorge Luis Guzmán.

FARC-EP (see figure 2.2). In some cases, the material conditions produced by war paradoxically preserved specific ecosystems, including the Amazon, in comparison to neighboring countries sharing the Amazon Basin.[22] To differing extents, certain areas were conserved due to the absence of state presence, and hence economic development initiatives, as well as the fluctuating and patchwork environmental conservation mechanisms enforced by FARC for ideological and militaristic reasons. In other moments, fronts of FARC were linked to illicit economic activities and other actions that, compounded with the operations of the public forces, destroyed and contaminated large swaths of territory and local ecosystems.

Since the official demobilization of FARC, reconfigurations of armed actors—labeled as dissidents, narcotrafficking, and criminal networks—have filled the power vacuum left in their wake. This has intensified socioenvironmental conflicts in the western Amazon and other regions. In 2019, Colombia was labeled the world's most dangerous country for environmental activists.[23] Of the 226 victims of killings recorded by a Bogotá-based NGO, Indepaz, between January and November of that year, 113 occurred in Amazonian states and neighboring Andean-Amazonian transitional zones

Rights of the Amazon in Cosmopolitical Worlds

with 80 percent related to land and natural resource disputes.[24] The human rights situation in the country has remained abysmal.

The notable upsurge in environmental crimes since the signing of the peace accords has led the state to increase the militarization of conservation efforts through the creation of burbujas ambientales (environmental bubbles). These bubbles are part of a national strategy to enhance monitoring, prevention, and control over deforestation, illegal mining, and the trafficking of fauna and flora through rearranged interinstitutional alliances between environmental authorities, police, military, and the district attorney's office. And yet, rates of deforestation increased even after the military-led Operation Artemisa launched following the issue of Sentence 4360.[25] In 2018, a forested area equivalent to twice the size of Bogotá was cleared.[26] The most recent bulletin on early detection of deforestation published by the Institute of Hydrology, Meteorology and Environmental Studies (IDEAM) in Bogotá revealed that 85 percent of the deforestation that occurred between October and December of 2019 occurred in the Amazon.[27]

Small- and medium-scale loggers are themselves struggling to navigate the swiftly changing legal, economic, and political landscape. An illegal logger attended an environmental education workshop that I cofacilitated during a sustainable development forum in Mayoyoque, Puerto Guzmán. He was a stocky man in his early fifties with weathered features and large calloused hands. When he stood up and confessed to the workshop participants that he cut down one hundred trees in a single day, tears welled up in his eyes. He frustratedly asked why the government failed to support viable transitional economic initiatives for rural families. He wanted more government initiatives for community-led conservation and reforestation of the Amazon rather than an increase in police and military blockades to arrest people involved in illegal timber trafficking. Militarized conservation tactics are one mode of perpetuating war by other means in official times of peace. Repressive conservation strategies continue to criminalize rural communities without offering them viable economic alternatives that are agroecologically appropriate to the Amazon. Environmental bubbles have not established permanent spaces for community participation in planning and policy decisions. The peace accords promised integral agrarian reform, territorial development, and the democratization of repressive antidrug policy, but the reality is far from this ideal. Furthermore, the militarization of conservation has not produced political guarantees and protection for the human rights of social leaders, environmental activists, and excombatants.

2.3 Oil pipeline blown up by armed groups. Image archive of *Semana* magazine.

In 2016, in anticipation of the signing of the peace accords, the president of the state oil company, Ecopetrol, was quoted in the newspaper affirming that "peace is going to permit us to extract more oil from the zones that were prohibited by war. ... With peace we hope to have the possibility to enter into Caquetá with much more force, Putumayo, Catatumbo, areas that were difficult to access."[28] Between 2004 and 2018, the national government signed sixty-seven contracts with nineteen companies for the exploration of oil reserves in the Caguán-Putumayo sedimentary basin of the Colombian Amazon (see figure 2.3).[29]

An artificial separation of soil from subsoil in resource-management logics disassembles the integral nature of the soil as a natural body. This divide also ignores the fact that any access to subsoil wealth must necessarily go through the soil and involves its occupants, guardians, and property owners. The 2014–18 National Development Plan created new administrative divisions on the surface of the territory that officially reclassified areas of Colombia's Amazon into two parts—the Center-South Region and Plains Region—opening the door to even more extractive industries. This change allows for more insidious modifications to the established use of soils, allowing new kinds of development projects and economic models to enter into the territory.[30] These reclassifications are compounded by the fact that the national government with the support of the courts has blocked the constitutional right of municipalities, and hence citizens, to protect the ecological and social patrimony of their territories and define the use of their soils through popular referendums and other participatory mechanisms.

The state is signing off concessions to multinational corporations for the development of mining-energy projects, while criminalizing rural communities for deforestation. This is a profound contradiction for rural communities living in the Amazon and deeply hinders environmental conservation efforts and the possibility of compliance with the court orders established in STC 4360. From the perspectives of these communities—be it coca growers whose livelihoods have been eradicated and poisoned by glyphosate during the ongoing war on drugs, Indigenous groups who have witnessed their territories reduced to a fraction of their ancestral lands, or social leaders facing death threats and forced displacement for defending the region's biodiversity and water sources from extractive projects—"the world's end" seems more of a repetitive, vicious cycle and permanent, latent possibility than a recently tangible threat that can be remedied by a court sentence.

Exclusions versus Intercultural Pacts and Legal Pluralism

On August 22, 2019, the Tribunal of Bogotá, in charge of monitoring compliance with the court orders dictated in Sentence 4360, declared that "none of the (government) entities have demonstrated full compliance, and on the contrary, are far from reducing deforestation to net zero in the Amazon." Indigenous communities who are custodians of 47.2 percent of the Basin responded by calling on the state for the second time to devise an intercultural pact. They argued that an intergenerational pact should not only be established between adults and youth, as was outlined in the sentence. They called for inclusion of their communities and the incorporation of their ancestral practices and autonomous environmental and territorial planning mechanisms. This included Indigenous jurisprudences and decision-making power as "public authorities" that can guarantee the cultural and territorial integrity of the Amazon.[31] The surrogate role that the children and youth plaintiffs assume for future generations within the structure of the sentence inadvertently assumes proxy not only for the region's Indigenous communities but for all of its rural and urban inhabitants. The call for an intercultural pact on the part of Indigenous communities renders visible a series of exclusions perpetuated by the legal sentence.

The centralized character of Sentence 4360 departs sharply from the earlier rights-of-nature case involving the Atrato River watershed. This case was driven by the concerns of local interethnic riverine communities, and resulted in a collective body of community guardians of the river who act on its legal behalf in planning for the watershed's decontamination and restoration.[32] One innovative aspect of the Atrato River sentence is that these orders can be interpreted and adjusted to social changes arising during implementation. Lawyer Felipe Clavijo Ospina characterizes this case as a dialogic rather than administrative sentence. Its effectiveness is progressive and involves a permanent participatory monitoring mechanism to ensure a processual implementation through the commission of river guardians.[33] The adoption of an innovative theory of biocultural rights permits the court to acknowledge the jurisdiction of Indigenous and Afro-descendent communities as authorities, guardians, and decision makers regarding the care of the river, in alignment with Indigenous Amazonian communities' call for an intercultural pact.[34]

The result of the exclusions built into the sentence of the Amazon manifested in five regional workshops held in Florencia, Mocoa, San José de Guaviare, and La Macarena between July and August 2018. These meetings were attended by a group of the young plaintiffs in the case and different representatives of social organizations and community members of the Amazon with the purpose of disseminating the sentence and discussing the creation of the Intergenerational Pact. The workshops were conducted in the principal cities of the Amazon, meaning that rural communities—the focal point of impacts of the court orders—were underrepresented.[35] One of the principal concerns expressed by attendees was the punitive quality of the sentence's initial repercussions. Local residents worried that state interventions to enforce its implementation would continue to punish them rather than invest in the socioenvironmental welfare of the Basin. They emphasized that forest conservation and reforestation are currently not viable economic options for sustaining their livelihoods. Severe obstacles prevent them from obtaining loans to support agroforestry projects. The existing credit system actually promotes deforestation because loans are almost exclusively granted to finance cattle ranching and agro-industrial projects.

Local governance and territorial ordinance issues were also raised during the dialogues. Many were outraged that the national government had committed to zero net deforestation in the Amazon but continued to grant environmental licenses for extractive mining-energy projects in the region. The state seemed mired in contradictions when it negated the forestry vocation of Amazonian soils while also assuming international and national commitments to safeguard these forests. With their visions for the territory excluded, Indigenous, Afro-descendent, and campesino communities feared that the national government would continue to introduce decontextualized projects that would negatively affect their ways of life. No less important were their worries about violent threats made against social leaders who organize their communities in defense of the environment.[36]

Residents also expressed concerns about whose interests were being represented and protected. Should extractive industries, as users (usuarios) of the territory, be involved in financing environmental planning and delimiting conservation zones? Beyond ordering the elaboration of an Intergenerational Pact, the court decreed that all municipalities of the region had to update their territorial planning documents within five months of its decision. This mandate did not respect the timeframes established by law to carry out territorial planning. Nor did it consider that the majority of the region's municipalities do not have risk management studies, currently a requirement for territorial planning.[37] Courts, judgments, and monitoring mechanisms can stimulate the formation of basic institutional capacities to deal with the structural problems that led to the emergence of litigations themselves.[38] However, the manner of acquiring these abilities in municipalities that lack infrastructure, resources, and technical capacities can perpetuate sectoral conflicts and economic dependencies on actors with financial capital and regional presence.

In 2018, Corpoamazonia signed an agreement with the National Hydrocarbon Agency and the Panamerican Foundation for Development to finance the technical assistance needed to update territorial planning and define zones of forest cover that should be conserved in fourteen municipalities of the Amazon.[39] Given the resources of the petroleum industry, it is not surprising that they were building alliances with a regional environmental authority. Conservation International Colombia, World Wildlife Fund, and other environmental NGOs operating in the Basin finance aspects of their projects through cooperative accords with the oil sector. Controversies over competing interests often intensify when governmental agencies are not allocated sufficient funds or personnel to support new legal responsibilities. Now that these regulatory and advocacy networks are involved in the implementation of newly recognized rights of nature, the question arises: how can these institutions work toward bettering the conflicts between incompatible interests and visions for the present and future of the region?

Propositional Reflections

When I interviewed Judge Tolosa in his chambers in Bogotá, I did not interpret his talk of "innovation" and transgenerational protections as a desire on his part for celebrity status or public appeal to a populist base. There is no doubt that Sentence 4360 is an important court ruling in its philosophical constitutional foundations and as an example of judicial activism in the promotion of biocultural rights and emergent climate-change jurisprudence. It has potential to act as a protective tool against the massive deforestation that is overtaking the Amazon, even as neighboring countries like Brazil engage in aggressive logging under the current Bolsonaro administration. The judge explained to me that the sentence was a "regional gamble" to inspire people, judges, and governments in other countries sharing the Basin to make similar demands and legal moves.

It is also important to recognize the potential embedded in the implementation and monitoring mechanisms for STC 4360 that the Tribunal of Bogotá is carrying out. This process is unique: the Tribunal has cited ninety-four entities in twelve hearings to request accountability and to redirect the implementation of the sentence. This is unprecedented in Colombia.[40] In these instances of reorientation, possibilities arise for dialogue and processes of supervision that can stimulate discussions over political alternatives to solve the structural problems detected in the ruling. But what would a dialogic approach to justice look like if it were to acknowledge the ontological diversity that is necessarily part of the situations leading to the origin of such a sentence and not only its implementation schemes?

Judge Tolosa revealed that he had little knowledge of the historical and contemporary realities of the region and its inhabitants or of the complex consequences of the court orders for local communities and regional authorities. He had not considered how the ruling might criminalize the same rural sectors long stigmatized, persecuted, and structurally excluded from spaces of decision-making in the territory. An enormous gap exists between normative frameworks and the complex realities of the worlds that judges or lawyers make decisions over. As Tolosa explained, "A judicial decision cannot resolve these kinds of problems. These are issues for public policy." This comment draws a false divide between judicial and political spheres where the former claims to simply produce legal rulings that are channeled to the latter for the crafting of legislation, regulatory protocols, and political decision-making with far-reaching public effects. Law is a product of sociopolitical realities, as Tolosa himself told me. The law should not be informed by a prescribed normative world that collapses or denies these complex realities. It is precisely these concrete and plural realities that should determine legal responses and not the other way around.

My intention is not to criticize this particular magistrate or to ignore the role of the courts in "structural sentences" that have taken place in the country over the last twenty years, but it is instead to make propositional observations regarding the imperative to train judges and lawyers in the historical analysis of territorial and socioenvironmental conflicts in the specific regions where they work or make interventions. Integral to this exercise is the creation of interdisciplinary teams that can support them in analyzing these contexts to inform subsequent legal decisions, which often exacerbate rather than ameliorate conflicts between different social groups and their particular worlding practices. Decisions by judges and legal concepts devised by attorneys far removed from local realities risk revictimizing and disregarding already marginalized populations with unequal access to justice mechanisms. A judicial decision without knowledge of the territory and of the realities of local communities through at least a judicial inspection or in-situ visit runs the risk of deciding blindly and with uncertain impacts.[41]

Multicultural constitutional reforms in the country's 1991 constitution introduced differential and asymmetric rights and social protections that aspire to guarantee the cultural integrity of Indigenous and Afro-descendent groups. These special protections have not been extended to campesino communities even though they share similar experiences of social exclusion and histories of agrarian-based violence.[42] The idea is not to fit campesinos into the same mold of current multicultural law with its promises and shortcomings. A more interesting question is how to repair fractured relations between rural communities that are not only a product of the social, political, and economic inequalities linked to decades of war but also of the divisive premises of multicultural constitutionalism. Equally important, how can we imagine and implement more capacious and plural practices of justice? Is it possible to recognize the particularities of Indigenous peoples' territorial relations and origins while also building on the diversity of rural realities and variations of justice existing in a given place?

Intercultural modes of territorial planning incorporate community-based practices of justice, which are more creative, situated, and pluralistic, and they can work alongside, even if in friction with, the normative frameworks of ordinary courts. There are 102 Indigenous groups in Colombia that have historically developed justicia propia, or their own principles for resolving internal problems. Through their Ley de Origin (Law of Origin) and Indigenous jurisprudence, these groups have strengthened their traditional authorities, while recovering forms of conviviality and autonomy.[43] While not yet officially recognized, Afro-Colombian community councils in collective territories also call for the recognition of their ancestral practices of justicia negra (Black law).[44] Communal-based justice is also part of many campesino organizations.

How would the legal rights of the Amazon have been structured differently if the legal ruling had been based on an intercultural pact produced through extensive dialogue with local communities and regional authorities? What proposals to mitigate deforestation would emerge if Indigenous communities, cattle ranchers, representatives of oil companies, loggers, and cocaleros (coca growers) were invited to the table in the copresence of jaguars, reemerging secondary forests, cattle, coca plants, and ancestral spirits? This scenario could be characterized as a space for cosmopolitical dialogue that offers the possibility of weaving unexpected alliances that might control the excesses produced by the systematic exclusion of local actors and their diverse lifeworlds. The idea is not that all regional actors would necessarily (or even want to) participate in spaces of conversation and debate, or that conflicts between different sectorial interests, visions for development, and life proposals would or should be completely resolved. Rather, a cosmopolitical proposal offers an alternative to the combination of law and order (i.e., security councils, military checkpoints, informational sessions, and prior consultation) that has not provided truly participatory opportunities and political guarantees for rural communities, nor dialogic spaces that attend to territories, forests, and watersheds as living beings and vital relations.

Repairing harms to forests and their spiritual guardians offers a completely different starting point to framing the problem in terms of crime and punishment. This does not mean that offenders should not be penalized or that environmental crimes and human rights violations are not occurring. However, financial sanctions have not stopped those who have the economic capacity to face an eventual case of penalty. They have not affected large deforesters or led to the capture of a single large-scale offender in the country's Amazon.[45] Local residents are tired of the tensions provoked by the actions of the public forces against the most vulnerable populations and the fact that state presence is reduced to police patrols and military checkpoints. It is necessary to take a step back to reconceive the problems at hand, and hence the solutions proposed, in order to avoid replicating historical exclusions at the root of ongoing territorial disputes. Concepts of harm and reparations can transcend conventional boundaries of administrative jurisdiction, categories of victimhood, and secular approaches that reinforce a modern division between nature and culture.

Indigenous jurisprudence and life philosophies have already informed more capacious and situated understandings of environmental harm and protection. For example, Colombia's 2011 Law of Victims for Indigenous Communities incorporated the notion of territory as victim, recognizing it as "a living whole and sustenance of identity and harmony" that "suffers damage when it is violated or desecrated by the internal armed conflict." In this law, "spiritual healing" is considered part of the integral reparation of a territory.[46] Rights *of* a territory are acknowledged rather than conventional proprietary framings of rights *over* it.[47] This kind of justice framework focuses on the rupture and potential repair of relations—socioecological as well as nonsecular relations that make and unmake a place and diverse lifeworlds. Addressing harms to territories permits a collective imagining of noncarceral alternatives to socioenvironmental damages. The focus on reparative and transformative justice practices can incorporate other-than-human actors into frameworks of damage, well-being, and repair. Transformative justice also attempts to address structural-level social issues, such as poverty and government or illegal armed actors' actions (or inactions) that have led to, permitted, or escalated socioenvironmental conflicts.

Politics can act to suppress the political—to reduce necessary ontological conflicts, passionate attachments, and diverse and dissenting worldviews and lifeworlds.[48] Bettering conflict entails creating opportunities to confront and live out disagreements. These disagreements may be related to the very conceptualization of the problems in question. They also involve recognizing the various realities at stake when the Amazon is deforested and degraded, including beings and relations extending beyond biologically informed concepts and modern environmental categories. Bettering conflict implies reorienting the expectation of resolving conflicts or merely experiencing them as dissensions that must be forcibly quelled through repressive strategies, be it the militarization of conservation or the managerial logics and corruption involved in the granting of licenses for natural resource exploitation.

Along similar lines, ordinary and criminal courts can inadvertently suppress pluralistic and place-based variations of justice. Like all environmental law, rights-of-nature cases must directly challenge the relentless drive toward economic expansion and unbridled exploitation of peoples and places. It is insufficient to merely attempt to mitigate its surfeits by allowing industrial-level extraction while criminalizing local residents.[49] The proliferation of sentences recognizing rights of nature cannot occur at the expense of excluding the voices of inhabitants who are born in, live in, die in, and defend the same territories and ecosystems that are being legally recognized and protected, only with their participation as an afterthought.

Indigenous communities, alongside Afro-Colombian, LGBT, environmental, feminist, agrarian, victim-led, and pro-peace movements, have all contributed to the diversity and extension of legal pluralism propelling contemporary rights-of-nature debates and juridical constructions. The idea is not only to fill rights-of-nature sentences with more or better substance and achieve effective compliance with court orders. It is equally important to ask whether the concept of rights and the extension of newly constructed rights are what particular territories, communities, and their diverse lifeworlds want or need.

Rather than a hurried and universal celebratory reaction to the proliferation of rights of nature, I ask that we also consider their unintended consequences, ethical complexities, and grounded particularities as well as the aspirational legal horizons that may be expanding in times of increased global reactions to climate change and the unbridled and catastrophic destruction of the Amazon Basin. Each case needs to be studied in its origins, trajectory, and imagined and dictated implementation schemes. This involves identifying the reproduction of absences, or excesses, at stake when environmental law attempts to confront its limitations through a biocentric turn. It also entails fostering dialogues between a complex range of regional actors where conflicts and disagreements over how to address the phenomena of deforestation, environmental degradation, and climate change can be lived out in capacious ways that question assumptions about who are relevant (better yet, existing) actors and how to formulate the problems that need to be solved.

Engaging with the violent realities and colonial legacies of massive deforestation and its connections to climate change requires creative and inclusive approaches that do not replicate historic exclusions. Rather, they must seek to "transform each (actor's) relations with his or her own knowledges, hopes, fears, and memories."[50] This, of course, also implicates legal-centric visions that assume to know the moral partitioning of right and wrong and categories of victimization in a society, including assumptions about the best way to represent and protect marginalized "others," such as forests, ancestors, and jaguars. The gathering of cattle ranchers, loggers, Indigenous, Afro-descendent, and campesino communities, urban residents, youth and elders, regional environmental and state authorities, NGO workers, lawyers and judges, illicit coca growers, illegal and ancestral miners, representatives of oil and mining corporations, police and military forces, and any number of other actors significant to the conversation, is only a first step. However, it most certainly should not be an addendum.

Local communities, regional actors, and their diverse territorial relations are at the core of a legal sentence that aspires to wrestle with the complex impacts of the deforestation of the Amazon, not because they are the only or even primary engines of deforestation and environmental degradation but precisely because it is their lives that are most directly implicated as well as their know-how and territorial presence that will guarantee (or not) any legal ruling's implementation and future success. The long-standing struggles of the region's inhabitants to create alternatives to extractive-based development models, militarized antidrug policy, and ongoing conditions of structural violence existed long before the symbolic influence and legally binding obligations of an innovative court ruling. It seems most instructive to take lessons from those who have not only survived but who continue to endure and flourish in the face of their "world's end" rather than to ignore the ancestral and popular roots of contemporary biocentric legal moves and expansion of right-based frameworks.

Notes

1. Serje, *El Revés de la Nación*.
2. Lyons, *Vital Decomposition*.
3. See Díaz Parra, *Entre el Estado unitario* for recent details about oil blocks granted in the Colombian Amazon.
4. De la Cadena, *Earth Beings*, 15.
5. De la Cadena, "Interview with Marisol de la Cadena."
6. Stengers, "Cosmopolitical Proposal," 996.
7. See Andrea Lozano Barragán, Victoria Alexandra Arenas Sanchéz, Jose Daniel y Felix Jeffry Rodríguez Peña y otros v Presidente de la República y otros (2018), Corte Supremo de Colombia, Sala de Casación Civil, STC 4360-2018.
8. For a further summary of STC 4360, see *ESCR-Net*, "STC 4360-2018."
9. MacPherson and Clavijo Ospina, "The Pluralism of River Rights." Although dictated in 2016, this court sentence was not released publicly until May 2017. See Corte Constitutional, "Sentencia T-622/16."
10. Kauffman and Martin, "Can Rights of Nature Make Development More Sustainable?"
11. Gómez-Rey, Vargas-Chaves, and Ibañez-Elam, "El caso de la Naturaleza."
12. See García Arbeláez, "Los jueces del fin del mundo." At the regional level, two

former governors, Camilo Romero (Nariño) and Carlos Amaya (Boyacá) signed a political pact in 2019 recognizing rights of nature in their respective states.

13 Prior consultation reflects the right, defined in law, of Indigenous and Afro-descendent communities to dialogue with the government to ensure their collective survival and autonomy. These communities have to be consulted and must approve final laws, policies, or projects that may affect them and their territories directly. See Rodríguez and González, "La Jurisdicción Especial Indígena y los Retos del Acceso a la Justicia Ambiental."
14 Centro Nacional de Memoria Histórica, *Narrativas de la Guerra*; Lyons, "¿Cómo sería la construcción de una paz territorial?"; Wilches Chaux, *Base ambiental para la paz*.
15 See Jurisdicción Especial Para la Paz, "Unidad de Investigación y Acusación de la JEP."
16 García Arbeláez, "Los jueces del fin del mundo."
17 Macfarlane, "Should This Tree Have the Same Rights as You?"
18 Ariza, Ramírez, and Vega, *Atlas Cultural de la Amazonia Colombiana*; Centro Nacional de Memoria Histórica, *Petróleo, Coca, Despojo Territorial y Organización Social en Putumayo*.
19 Whyte, "Settler Colonialism, Ecology, and Environmental InJustice."
20 Estupiñán Achury, "Neoconstitucionalismo ambiental y derechos de la Naturaleza."
21 Taussig, *Shamanism, Colonialism and the Wild Man*; Ramírez, *Entre el Estado y la Guerrilla*.
22 Rodríguez, Rodríguez, and Durán, *La paz ambiental*.
23 According to the Irish advocacy group Front Line Defenders, more than three hundred human rights leaders were killed in thirty-one countries in 2019, and nearly half of those killed were targeted specifically because of their environmental activism. Of these deaths, 108 were documented in Colombia. See Tomassoni, "Colombia was the deadliest place."
24 El Tiempo, "En tres meses, 120 líderes sociales han sido asesinados en Colombia."
25 I thank my colleague Carlos Olaya for sharing information about the Campaña Artemisa through a personal communication on November 11, 2020. See Presidencia de la republica, "Con la puesta en marcha de la Campaña Artemisa."
26 Armenteras, "¿En qué quedó el medio ambiente?"
27 Infoamazonia, "Se disparó la deforestación en la Amazonia colombiana (otra vez)."
28 "La paz nos va a permitir sacar más petróleo de zonas vedadas por el conflicto."
29 Semana, "Putumayo, clave para el futuro petrolero del país."
30 Díaz Parra and Aguilar Herrera, *Ordenamiento territorial y ambiental de la Amazonía colombiana en el posconflicto*.
31 Infoamazonia, "Indígenas dicen que la sentencia que otorga derechos a la Amazonia los deja por fuera." The omission of many Indigenous communities of the Amazon occurred even when six Indigenous authorities (ACIMA, AIPEA, PANI, ACAIPA, ACIYA, and ACIYAVA) are acknowledged as having participated in the

original action of tutela filed in the court (8). See also Dejusticia, "The Colombian Government Has Failed."
32 See Corte Constitutional, "Sentencia T-622/16."
33 Garzón, "Los Derechos de la Naturaleza se Sintonizan con la Conciencia Ambiental de Nuestro Tiempo."
34 Macpherson, Ventura, and Clavijo Ospina, "Constitutional Law, Ecosystems, and Indigenous Peoples in Colombia."
35 Dejusticia, "Colombian Government Has Failed."
36 This information was obtained through a personal email with Dejusticia on August 3, 2019, and an internal document submitted by Dejusticia to the Ministry of Environment and Sustainable Development on August 2, 2018.
37 Díaz Parra and Aguilar Herrera, *Ordenamiento territorial y ambiental de la Amazonía colombiana en el posconflicto*.
38 Rodríguez Garavita, and Rodríguez Franco, *Juicio a la exclusión*.
39 Corpoamazonia, "Catorce municipios del Sur de la Amazonia ya cuentan con Determinantes Ambientales definidas y actualizadas."
40 I thank my colleague Carlos Olaya for sharing information about the implementation and monitoring process for STC 4360 via personal communication on November 11, 2020.
41 I thank my colleague Felipe Clavija Ospino for sharing his perspectives on this point.
42 Duarte, *Desencuentros Territoriales*.
43 Rodríguez and González, "La jurisdicción especial indígena y los retos del acceso a la justicia ambiental."
44 Izquierdo, "Justicia negra."
45 Botero, "Señor Minambiente hablemos de deforestación en serio."
46 For more details on this law, see Decree-Law 4633 of 2011.
47 Izquierdo and Viaene, "Decolonizing Transitional Justice"; Ruiz Serna, "El territorio como víctima."
48 Rancière, *On the Shores of Politics*.
49 Gonzalez, "Bridging the North-South Divide."
50 Stengers, "The Cosmopolitical Proposal," 1002.

3

"We Are Not Pests"

Alyssa Paredes

3.1 (*previous page*) Original drawing by Feifei Zhou.

IN 2005, a contentious environmental campaign exploded across the plantation-dominated landscape of Mindanao, the southernmost region of the Philippines. Brought together under the civil action group Mamamayan Ayaw sa Aerial Spray (Citizens against Aerial Spray, henceforth MAAS), locals from the rural outskirts of Davao City, in Mindanao, raised fists and placards in protest against chemical drift from crop dusters, small aircrafts used for the aerial application of fungicides on nearby banana plantations. Invisible, uncontainable, and uncaptured by modern managerial schemas, the chemical fallout is carried by the wind beyond farmlands onto houses, home gardens, schools, public roads, and water sources. Residents along plantation peripheries find white and yellow specks on their clothes and in their laundry, on the fruits and vegetables they grow in their yards, and in the water drums they use for drinking, cooking, and doing the dishes. Plantations rely on the fungicides primarily to control *Mycosphaerella fijiensis*, a pathological fungus responsible for the airborne, leaf-spotting disease Black Sigatoka. But consequences beyond the target fungus are undoubtable, as aerial spraying's blanketing effect does not discriminate between microbial, plant, animal, or human life. Demonstrators with MAAS have denounced their entanglement in such a system that sees their bodies as subhuman, a legacy of the colonial capitalist encounter that renders them undifferentiated from the nonhuman Others with whom they find themselves caught in the chemical haze. In a campaign that rose from the streets and courts of Davao City to the Supreme Court between 2005 and 2016, they have adopted the provocative rallying cries, "I am not a banana!" and "We are not pests!" (Visayan: Dili mi peste; see figure 3.2).

3.2 "We Are Not Pests! Stop Aerial Spray." Mamamayan Ayaw sa Aerial Spray Protest at Davao City Hall, August 26, 2006. Photograph courtesy of Interface Development Interventions for Sustainability.

The slogans of MAAS strike a different tone than some other antitoxic campaigns that might be familiar to people in the Global North. Since the publication of Rachel Carson's exposé on DDT, the world has seen a proliferation of accounts on the hidden dangers of agrochemical biocides. These literal "life killers" collapse the boundaries between biological cultivation and annihilation, and in their toxic persistence they deny the possibility of future flourishings in a dynamic Deborah Bird Rose describes as "double death."[1] While many of these lethal chemicals remain in regular use around the world, lively civil movements have risen up against them, most commonly by calling for bans to certain toxicants. Across Europe, civil organizations have demanded restrictions on chlorpyrifos and glyphosate; in Central America, on DBCP (1,2-Dibromo-3-chloropropane); in Kerala in India, on endosulfan; in Dalian in China, on paraxylene; and in Bangkok in Thailand, on paraquat, to name only a few. In contrast, the slogan "Dili mi peste" shifts the rhetoric of protest ever so slightly away from particular chemical contents and toward a subjective state, one that desires human-nonhuman boundary drawing at a moment of compromised human dignity. Such a rallying cry reflects an important difference in the experience of chemical pollution and expresses a particular vision of justice. However, it also has an alternative reading that complicates multispecies justice: it insists on anthropocentric rather than collective ecological justice. For what are the slogans "I Am Not a Banana!" and "We Are Not Pests!" but concessions to spray nonhuman beings so long as humans are left unharmed?

This chapter explores how such a vocabulary of protest emerged as locals grappled with conditions of possibility and impossibility deeply shaped by interspecies intra-actions in and around the industrial plantation. Rather than dismiss the movement for putative anthropocentrism, it recognizes activists' strategies as an ethics of exclusion following Eva Haifa Giraud, who states that exclusion need not be synonymous with processes of marginalization and oppression, nor simply imply negation. On the contrary, exclusion plays a creative and constitutive role as a political intervention and a site of accountability.[2] Giraud's call comes at the helm of recent theoretical developments focused on ethics based on relationality and entanglement, which "struggles to accommodate things that are resistant to being in relation."[3] Such challenges to entanglement animate politics in and around plantation settings, where Sophie Chao, for example, has described how certain crop species are unable and unwilling to participate in reciprocal relations of coexistence.[4] Certain entangled realities foreclose the possibilities of other ways of being; thus, an ethics of exclusion calls attention to the practical materialities of more-than-human entanglement, especially those that limit political action. Taking this as a starting point for understanding the MAAS campaign, I argue that when operating within limited legal frameworks and delimiting capitalist logics, justice can often become an exclusionary project, and species divides the fault lines along which those exclusions are understood, organized, and contested. What results are competing claims over inclusion in the ambit of justice.[5] For many people, this is a matter of course when the divides between the human and nonhuman and between Life and Non-life—distinctions taken to be natural—grow ever less distinct. Chemicals, prime technologies for the control and containment of microscopic life, render tenuous their use against pests and their use against humans *as* pests.[6] Such precarious boundaries are a matter of life and death, as they render as up for grabs the question of who or what in the social hierarchy of beings deserves protected life. These are everyday points of contention on the plantation, one of modernity's prime institutions for deciding which forms of species life are made to live or let to die, and in whose name.[7]

Local residents who have joined MAAS's anti-aerial spray movement experience acutely the dark biopolitical logics that degrade their sense of humanness. Living on the perimeter of industrial farm properties where multinational corporations grow Cavendish bananas for export, they belong to residential communities that are emblematic of agrarian life in Mindanao. There, an estimated 500,000 hectares of prime agricultural lands have been dedicated to growing export rubber, sugarcane, bananas, pineapples, oil palm, coconuts, and cacao, each with varying chemical demands and all interspersed in patchwork fashion into the rural landscape. Of the total hectarage, 68,000 hectares are dedicated to the cultivation of Cavendish, the sole banana variety for export to major markets in Japan, China, Korea, the Middle East, and elsewhere. With approximately thirty liters of fungicidal mixture applied for each hectare of aerially sprayed banana land, and likely greater volumes for those that are treated with ground-based methods, several million liters are released onto Mindanao with every spraying. Documentation on spraying cycles is not centralized, and practices differ between plantations and between seasons, but on many industrial banana farms, crop-dusting occurs as frequently as twice a week.

My chronicle of locals' experiences is one aspect of a larger project, based on two years of multisited fieldwork in the Philippines and Japan. Here I draw on the first half of my fieldwork, which was conducted on Mindanao's banana plantation belt in the provinces of Davao del Sur, Davao del Norte, and Cotabato. I carried out immersive research in the residential areas, home gardens, and roadsides that border industrial farms, spaces that elide the knowledge and practical concern of corporate managers. These offered critical ethnographic counterpoints to the research I was also doing in the offices of top management, growers' cooperatives, plant nurseries, packinghouses, and plantation grounds. In fieldwork on either side of plantation boundaries, I was guided by an interest in what was included and excluded from the industry's production calculus: What factors, both social and environmental, were accounted for, and what were not? Through what technical, bureaucratic, or political structures did such processes of exclusion become not only possible but also standard operating procedure? And what routes to justice have locals charted in response? At no point did this dynamic become more apparent than when thinking about chemical diffusion. I did research on the MAAS campaign between 2016 and 2018, ten years after the campaign began and right at the cusp of its conclusion with the Supreme Court decision in 2016, using a combination of participant observation, interviews, and a review of government documents and media items. I worked most closely with MAAS's central cohort of activists and two of their closest NGO allies, the environmental advocacy group Interfacing Development Interventions for Sustainability (IDIS) and the alternative legal resource group Sentro ng Alternatibong Lingap Panligal (SALIGAN). Our conversations revolved around how and why they chose to frame their campaigns in the ways that they did, why their particular language of protest had the kind of purchase that it does, and what visions of a just future it entailed. It soon became clear that the answers to all these questions involved negotiations within and across species boundaries.

I begin with an introduction to the plantation landscape, bridging the dynamics between animal sentinels and the nexus of virulent fungi and chemical compounds with the formation of MAAS's protest rhetoric. I discuss how subjection to the state of animality or objecthood, which the campaign rallied against, builds on longer legacies of colonialism at work in Mindanao's plantation landscape and more broadly. Approaches to justice by MAAS would have implications in and out of court, as I then demonstrate, as they ran up against scientific regulatory paradigms. The 2016 decision from the highest tribunal subsequently raises discussions on the disparate and constantly vacillating values attached to human and animal life under capitalist production systems. I conclude with reflections on interspecies competitions for inclusion in the project of justice, offering an alternate vision by reclaiming the figure of the so-called pest.

"The Start of Us Making Noise"

Dagohoy Magaway lives in a home tucked away in Tugbok district, a rural residential area of Davao City. I first met him at a café in the city, then multiple times after on the veranda of his house. A kind-looking man who wears glasses hanging off a brown string around his neck, Dagohoy comes from a lineage of activists and is named after the revolutionary who led the longest rebellion in Philippine history, an intriguing foreshadowing of the ten-year campaign he would lead against aerial spraying with MAAS. Before he became one of the most outspoken public figures in the movement, he was primarily a coconut farmer who sold rice cakes to plantation laborers at the company canteen. Three times a week, Dagohoy fell under the chemical showers while on his morning commute. It hurt his eyes and made his skin feel sensitive, such that he could not wash his face or take a proper shower for days after.

Banana plantations expanded into Dagohoy's neighborhood along the gently sloping piedmont to Mt. Apo in the early 2000s. However, the industrial agricultural estates are a legacy of export-oriented agricultural intensification that push back through time: from the Spanish and American colonial sugarcane hacienda enclaves of the 1850s, to the Japanese-dominated Manila hemp plantations of the pre-war period, to the American postcolonial expansions in tropical fruit agribusiness that continued well after the formal end of US occupation (1898–1946). The beginnings of Philippine export of Cavendish bananas in particular can be traced back to the 1950s, when the United Fruit Company, facing the problems of land expropriation and crop disease, turned to the Southeast Asian country as a potential site for diversifying its investments.[8] The Philippine government under President Carlos P. Garcia welcomed United Fruit's proposal in 1958 as part of new policies to increase dollar earnings through export-oriented agricultural industrialization.[9] For their purposes, stakeholders eyed the then-typhoon-free Mindanao, which was undergoing a government-assisted influx of internal settlers from the North and Central Philippines, largely displacing Indigenous tribes and their lifeways. With the participation of settlers turned growers, plantation agriculture for bananas in the style of the New World was transposed to the island beginning in and radiating out of the provinces of Davao del Norte and South Cotabato, both sandwiching Davao City, in 1967.[10] Driving these developments were the multinational corporations that continue to dominate the industry until this day, including, in the order they appeared, the Standard Fruit and Steamship Company of New Orleans (subsequently Castle and Cooke, now Dole), the Del Monte Corporation of San Francisco, the United Fruit Company of Boston (now Chiquita), and, entering two decades later, Sumitomo Corporation of Japan. Accompanying these firms were intensified ecological changes to ecosystems already transformed by internal settlement, as well as the conventions of agricultural science, techniques of crop management, and chemical regulation adopted from the Western hemisphere as the primary means for understanding and managing those changes. The aerial spraying of fungicides using crop dusters was part and parcel of these developments, and while the precise history of its introduction is unclear, it was a method used on Philippine banana plantations by the late 1970s if not earlier.[11] Today, the multinationals' local subsidiaries, Dole-Stanfilco, Sumifru Philippines, Chiquita-Unifrutti Tropical Philippines, and Del Monte Philippines, all export bananas that have been treated with aerial spray.

In the first several months that Dagohoy spent living under the sprays, the effects on plants and animals around the village were the first to become visible. He recounted the story of the plantations' arrival in the early 2000s, and of his gradual awakening to the threats of the drift in multispecies terms. "When they cleared the [land for the] farms, they bulldozed trees to the side. In nature, the stage of decomposition, it turns out, beetles find that delicious! It became the home of the beetles, those decomposed trees." He spoke specifically of the Coconut Rhinoceros Beetle, a parasitic and invasive pest that bores into treetops and feeds on the sap, leaving the leaves with distinctly jagged, zig-zagging cuts.[12] Aerial spraying made things worse. He and his family tended to two small coconut farms on land that had been saved from the clearings wrought by plantation expansion. Both family farms are in areas that bordered banana plantations that were relatively small, irregularly shaped, and noncontiguous. As coconut trees tend to be taller than banana plants, they are often within immediate range of the crop duster drift. Once the sprays fall onto trunks and are heated by the sun, they burn the bark. The decay emits a smell that attracts the beetle. The issue was compounded by the fact that the fungicidal sprays tended to kill off not only the targeted *Mycosphaerella fijiensis* fungus but other natural fungal antagonists that had served to control the beetles' spread. In Dagohoy's words, "Because they spray fungicide, it doesn't just kill Sigatoka, it kills all fungus [sic]. Because the natural enemy of the beetle has disappeared, [there is] no more control for the population, so it [the beetle population] really grew! Just imagine, in our farm, after two years, our coconut production decreased by 50 percent because of the beetles. When the leaves of your plant [are unhealthy], you cannot expect to have a lot of harvest. After all, the plant's kitchen [the leaves] is ravaged [Tagalog: Eh kung sira ung nagluluto ng pagkain] . . ."

As his coconut dwindled, Dagohoy saw his income drop by a half, a trend that many of his neighbors experienced as well. "This was the start of us making noise," he recounted.

It is no coincidence that the coconut farmer would eventually come to grasp the potential long-term impacts of the sprays on his own body by first watching the changes in the bodies of plants and animals. Anyone who has found themselves under the chemical mist knows the profound sense of disorientation that accompanies the experience. You hear the doppler effect of the crop duster's roaring engine, and the subtle feathery sound of descending mist. You watch as the milky-colored drift leaves the valves of the crop dusters overhead, but the drift dissipates instantaneously, going everywhere and nowhere at once. Then there is nothing you can see, nothing you can feel, and nothing you can smell. It is an experience of the "chemical sublime," taking from Nicholas Shapiro after Kant, a dulling of the perceptual faculties matched only by a hyperawareness of the fragility of your own embodiment.[13] For a brief second, you are relieved and think, Did I manage to escape it? But before long the failures of your senses give way, slowly but forcefully, to a creeping sensation that colonizes your skull. You cannot shake it off; it accompanies you all day. You start suspecting that the toxic drift is everywhere. You cannot quite explain what is going on in your body. Like Celia Lowe's notion of the "viral cloud," the drift reassembles and reassorts your body in ways that seem nonrational and nonrationalizable to you.[14] All you know with certainty is that it feels like an infringement.

Plantation Sentinels and Shields

It took only one experience under the sprays for me to decide that I would never return to the plantation for fieldwork on a morning of a cycle. However, there are several communities who watch the crop dusters from behind windows or under the eaves of wooden houses as frequently as twice a week. This includes women from the neighboring villages, who shared in Dagohoy's qualms and joined him in becoming central figures of the campaign. I most frequently met with a core group of five warm and outspoken ladies who shared stories and documents, doodled over in crayon by their children, as we chatted in home gardens only a few meters away from the nearby Cavendish banana plantations. Where these folks lived, areas separated from the main thoroughfare of the city's urban center, people had only the bodies of dead or dying plants and animals to see the chemical drift made manifest in its transubstantiated form. One woman, whom I call Geronima, explained to me how you could tell that the sprays had gotten inside when the cacao trees stopped bearing fruit, the moringa plants curled, and the chicken in the yard dropped dead. I began to think of chickens on the plantation as I would canaries in a coal mine.

Other species—birds, fish, mice, rabbits, guinea pigs, dogs, cows, even lichen—have long served as "sentinel devices" for humans, as "early warning systems" for the assessment and management of environmental pollution and risk.[15] While the notion of the sentinel has its roots in military surveillance, scholars have sought to reclaim the concept to refer instead to alternative—that is, "postcolonial and nonsecular"—modes of perception.[16] In serving as a warning signal of impending threat, they claim, sentinels also signal the existence and transcorporeality of a more-than-human collective, one that is marked as much by solidarity as by vulnerability.[17] Just as migratory birds with influenza might sound the alarm bells of an impending H5N1 pandemic within the human community, dead chickens on the plantation's peripheries trigger forms of more-than-human bodily reasoning. For When Dagohoy, Geronima, and their neighbors raised the cases of perishing livestock, I sensed that their concern was not only over the economic repercussions that would surely ensue but also over the uncertainty about whether human and animal bodies were *different enough* to ward off the same bodily effects wrought by a shared toxic environment. This was the obverse effect of Timothy Choy's "conspiracy of breathers," a notion inspired by the Latin root *con+spirare* meant to capture the sense of commitment to "breathing together" even while not "breathing the same."[18] Rather than sparking new solidarities in resistance to chemical pollution, locals found themselves anxious about the integrity of divisions between the human, animal, and plant worlds in the cloud of fungicidal haze.

Unable to escape from the chemical showers even in the shelter of their own homes, local residents have called upon animals and plants to serve not only as bellwethers of harm but as physical barriers from it. As a stipulation on Environmental Compliance Certificates, all banana plantations are required by the Department of Environment and Natural Resources to install thirty-meter-wide buffer zones of perennial foliage to protect public and private properties from the activities on plantation grounds. In reality, however, only a handful of corporations fulfill this legal obligation as buffer zones reduce planting area and hence income, as scholars have analyzed elsewhere in the world.[19] Because government representatives are lax in the monitoring and enforcement of their own regulations, locals are forced to contend with the consequences or otherwise find their own solutions. Geronima, for one, had planted a perennial hedge around her property so that other organisms might serve as her shield.

More-Than-Human Dynamics and the Language of Protest

Plantation sentinels hint at the profound sense of unease that comes when the boundaries between species are rendered mutable, something that Davao's locals turned activists would continue to confront throughout their campaign.[20] At the source of those boundary-bending processes are the dynamic intra-actions between plantation pathogens and chemical technologies of containment. Anthropologists and allied scholars have recognized that animal metaphors inform "organizational logics of action,"[21] and that chemicals "catalyze political projects."[22] The MAAS story draws closer to such analyses by demonstrating the formative effect of more-than-human dynamics on the particular styles and strategies of a social and environmental movement, a language of justice that has had legal consequences.[23]

Historical and ethnographic work has shown how environmental landscapes are never passive canvasses onto which humans draw out their mental designs. Invasive pests, pathogens, and epidemics shape the destiny of human-driven projects, and once empowered by those projects, they break out of the infrastructures in which they were intended to operate.[24] This was certainly the case on Philippine banana plantations, where the Sigatoka fungus has been transformed into a most formidable foe by the same conventions of industrial agriculture that have sought to control it. The tight monocropping of the Cavendish variety, while par for the course in economies of scale, turns the plantation into a playground for pathogenic fungi. Transferring from one plant to another with ease, these pathogens are also assisted by wind blowing over the plantation, unhindered by the tropical forest foliage that was cleared for the purposes of industrial cultivation. To make matters graver, Cavendish bananas, virtually the world's sole variety for export, are genetically homogeneous and thus practically defenseless to disease. Sarah Besky's reference to the "sick landscape" of the plantation, and to sickness as that which allows plantations to persist unchallenged, is particularly apt.[25]

Chemical pesticides have long been the go-to strategy for containing and controlling Black Sigatoka on industrial scales. However, everywhere that the pathogen has emerged, commonly used fungicides have decreased in their efficacy over a number of years, a result of the Sigatoka fungus's ability to adapt to synthetic chemicals and acquire resistance. At first, the Philippine banana industry responded by increasing the frequency of spraying schedules,[26] but when they realized that was only helping the fungi develop resistance, they came up with a more sophisticated strategy. They introduced something called the "chemical cocktail," a mixture of several different fungicides rotated semi-randomly.[27] One company, which has chosen to remain anonymous, shared with me the information that it alternates some eighteen different Sigatoka control fungicides in the cocktails loaded into its crop dusters. Of the eighteen, several have been identified by the California-based Pesticide Action Network (PAN) as "Bad Actor Chemicals" for possessing qualities that make them either highly acutely toxic to people, a cholinesterase inhibitor,[28] a known/probable carcinogen, a known groundwater pollutant, or a known reproductive or developmental toxicant. They are mancozeb (brand names: Biozeb, Dithane, Ever Mancozeb, Ivazeb), chlorothalonil (Daconil), propiconazole (Tilt), epoxiconazole (Opal), propineb (Antracol), and thiram (Banguard).[29] A vast majority of these fungicides are developed by the largest players in the agrochemical industry, Syngenta, Bayer Crop Science, Dow Chemical Company, and BASF, the "grand-kin" of toxic relations that extend the industrializing, racializing, marginalizing, and dispossessing logics of a capitalist "chemical regime of living" ad infinitum.[30]

Any one of these fungicides would tell an alarming "imploded history," to borrow from Joseph Dumit,[31] of toxic pervasiveness and persistence in organisms and in the environment through time.[32] However, to separate each out individually—as discrete entities or isolated molecules in the functionalist ways that industries have conceptualized them[33]—belies the yet more pernicious reality of their "intra-actions" as mixtures, which render the boundaries between one chemical and the next moot.[34] When applied not as solo chemicals but in a rotating concoction, the fungicidal cocktails effectively confuse the Sigatoka fungus and prevent it from developing a tolerance to any one agrochemical.[35] But beyond their immediate agricultural function, however, the cocktails also serve a very important political purpose. They make it impossible to parse out one chemical from another and thus disable the public from knowing what exactly is being sprayed when. When I asked one woman who had worked in a packinghouse for a Japanese company what she knew about the cocktails, she answered simply, "Only the supervisor and the mixer know what's in this, and the chemical content changes all the time."

Without knowing what those individual active ingredients were, Davao locals had no way of figuring out what was causing what. When activists first conceived of the MAAS campaign in those few years after aerial spray started in Davao, it was clear that theirs would have to be a movement about defending human dignity against the agricultural method of aerial spraying itself regardless of the content loaded into them. One writer named Stella Estremera elucidated this forcefully in an op-ed article entitled "I Am Not a Banana," and she deserves to be quoted at length.

> Detergents and soaps have chemicals. They need to have chemicals to remove stains. We dip our hands into them when we wash the dishes, and they cannot kill us unless we drink them. But if I go around lugging a tub of soapy water and drench a passer-by against his will, then I'll get into trouble. Not because the soapy water was toxic, but because the passer-by didn't want to be drenched. Saying, "That's ok, friend, it's not poisonous after all" will not get me out of trouble.
>
> Banana pesticides have chemicals designed to kill *banana pests*. They were formulated to spray on bananas. Now the question: ... Why are

some people insisting that they can spray pesticides on *people* who never asked to be sprayed even with water? Simplistic, yes, ... it's all about my right to say, "I do not want to be sprayed."[36]

As Estremera's statement makes clear, MAAS was not protesting toxicity per se. Neither was it pushing for tighter chemical regulation. Instead, it protested a compromised subhuman state, the state of being undifferentiated from cheap fruit for devouring ("I Am Not a Banana") or pests for eradicating. It was propelled by a desire for human dignity founded on what might be termed "molecular sovereignty," an homage to the notion of molecular privacy, first coined by Robert van den Bosch, the eminent entomologist among those who developed Integrated Pest Management.[37] Van den Bosch's concept refers to the right of an individual to take a precautionary approach against exposure to molecules, whether harmful or benign, without requiring damage to be proven. While resonant, the term's emphasis on privacy overlooks the fact that what MAAS activists seek is less a perfect quarantine than a desire for sovereignty—that is, self-determination over the chemical conditions of possibility and, following Stacey Langwick, habitability.[38] But this vision of sovereignty was reserved for individuals who were inevitably human; it implied an ethics of exclusion that demarcated "banana pests," the intended targets of chemical application, from "people who have never asked to be sprayed even with water."

The Colonial Heritage of Interspecies Competitions for Justice

That MAAS activists would frame their campaign around an insistence on fundamental species demarcations points to the precariousness of the human in the more-than-human in the first place. This has long been a subject of postcolonial, Indigenous, queer, and Black feminist scholarship, which has brought attention to how the constructed nature of the human subject can often be contingent on the dehumanization of racialized colonial subjects, reduced to the state of animality or objecthood.[39] Philippine colonial history is laden with examples, such as young Filipino girls being put on display at a Coney Island zoo and circulating photographs of Filipino men labeled as "most monkey-like people in the world" when set alongside US colonial administrators (see figure 3.3).[40] Geronima herself has lineage in the Bagobo-Guianga tribe, one of the Indigenous groups whose members were brought to the 1904 St. Louis World Fair and placed in the "Philippine Reservation," the largest human exhibit on site.[41]

3.3 Photograph by Dean C. Worcester, labeled "Negrito man, type 1, and myself, to show relative size," Mariveles, Bataan (1901). Reproduced with the permission of the University of Michigan Museum of Anthropology, UMMAA 01-A-002.

"I am not a banana!" and "We are not pests!" allude to how colonial configurations are manifested, articulated, and resisted through food and animal metaphors. By rejecting the positionality of a food item for consumption, MAAS campaigners gesture at the divide between *who eats* and *who is eaten*. This is well known to scholars analyzing the intersection of food, race, and coloniality, who argue that eating is a boundary-drawing moment that marks the difference between the human privilege of the eater and the objectification of the eaten. A most well-known example is the trope of "Black edibility," or the eroticization and exoticization of Black bodies *as food*, a form of racist violence, rendered mundane, that pervades commercial markets familiar to consumers around the globe.[42] Bottles of dark pancake syrup, eliciting childhood nostalgia, come in the shape of a woman's body; and the popular French chocolate milk drink Banania is stamped with a Senegalese face locked into a perpetually obsequious smile, saying "y'a bon Banania," translated as "sho' good eatin."[43] It is impossible, too, to forget "Strange Fruit," the infamous song performed by Billie Holiday on the bitter crop that hangs from poplar trees of the Jim Crow South. For Dagohoy, the feeling of being undifferentiated from the commodities whose cultivation had caused him such anguish was always accompanied by a sense of entanglement in a world-political economy that is profoundly unequal. "That's what's happening to us just to feed those Japanese at our expense," he said to me once. "Just to feed another country, we have to pay with our lives" (Tagalog: Mapakain lang ang ibang bansa, kapalit na ng buhay namin).

Animal metaphors and, in particular, reference to the pestiferous in "We are not pests!" illuminate chemicals as striking and gruesome examples of colonial conflations at play. Pesticides in the African colonial context were used both to exterminate vermin *and* guerrilla fighters deemed "vermin beings" threatening to the white race.[44] During World War II, the chemicals that German companies used in the concentration camps of the Holocaust were purchased from the United States, where they had been applied for the control of both lice and the Mexican migrants considered to be their carriers.[45] Still today, dangerous interspecies conflations pervade popular media. Miriam Ticktin, for instance, notes how the metaphor of invasion has been used both for pests and pathogens and for human migrants and refugees, compelled by choice or circumstance to migrate to places where they are seen as not belonging.[46] She warns that the fear of intrusion, whether it is into the physical body or into the body politic, lurks nearby whenever bodies are thought of and experienced as autonomous and self-contained. When metaphors overlap, they have serious material consequences across species divides. We see this clearly when British TV personalities dub refugees from Syria and Northern Africa as a swarm of cockroaches, and the containment of pathogens—like Zika and Ebola—becomes synonymous in practice with the containment of certain peoples.[47]

In response, several anthropologists have championed cosmopolitan impurity, plural interdependence, and an embrace of contaminated worlds.[48] They build on the work of Donna Haraway in her call for moving beyond modernist dreams of species purification.[49] In the realm of the chemical, Vanessa Agard-Jones has offered the concept of "kin/esthesia" to refer to the ways by which people's awareness of their body burdens contributes to their articulation of kinship with communities of chemical injury beyond multiple boundaries and scales.[50] These are important theoretical interventions where cultures of anthropocentrism and human exceptionalism, matched with unfettered capitalism, galvanize planetary destruction. However, they also raise the question of how to move past the attachment to notions of purity and the violence it engenders without denying subjects the ability for ethical judgment. After all, when the Other is a harmful toxin, what would it actually mean to speak of kinship or hospitality to people like Dagohoy and Geronima, who long for the pleasures of an uncontaminated environment and desire the sovereignty to define and defend their human bodily boundaries—in other words, an ethics of exclusion?

Modes of Protest and Regulatory Clashes

The particular mode of argumentation by MAAS, insisting on human dignity and molecular sovereignty over toxicant identification and control, had purchase among communities on the plantation's border. After its inception in 2005, the campaign expanded to include hundreds of members from nine villages (barangay) around the city.[51] Community organizers found important allies in Davao-based NGOs, most notably, the environmental conservation organization Interface Development Interventions for Sustainability (IDIS) and the Indigenous peoples' health advocacy group Kalusugan Alang sa Bayan, Inc. Supported by these organizations, MAAS took to the streets fronting Davao City Hall in mid-2006 after initial negotiations with banana companies were unfruitful. Once a weekly endeavor, the protests soon snowballed from thirty individuals to thirty jeeps-full.

After several months of campaigning, joint fact-finding missions, medical testimonies, and summary reports to the Sangguniang Panlungsod, the movement was met with some success. Then-mayor of Davao City Rodrigo

Duterte signed City Ordinance No. 0309-07 banning aerial spraying on February 9, 2007. The resolution behind the ordinance highlighted the importance of adopting a "precautionary principle" in policy, even in situations where detrimental health effects to residents, nearby watersheds, and agricultural businesses may not be firmly evidenced. In the hopes of enacting swift measures to alleviate local suffering, the ordinance allotted three months for companies to shift to ground and manual spraying before the ban's full enforcement. The turnover was quicker than institutions had anticipated. The Davao Medical Society, in contrast, had pushed for an aerial spray phaseout in one year's time. The Pilipino Banana Growers and Exporters Association, a group of the most powerful bananeros in the country, demanded no less than twenty-five years.[52]

The historic ordinance, the first of its kind in an area with a strong lobby from banana businesses, set off a legal battle that would seesaw through courts at all levels of government over the next ten years.[53] The growers and exporters association was quick to retaliate, and in media reporting, it not only denied the reality of drift, but they also insisted that aerial spraying was the "safest, most effective, and most accurate method" of chemical application.[54] It further denounced the use of the term *chemical cocktails* and suggested that the solutions were better characterized as *water cocktails*.[55] The chemicals and their doses were recognized by the Philippines Fertilizer and Pesticide Authority, hammering down the reality that, technically speaking, banana corporations had not been breaking any rules. Industry leaders asserted that, in an industry that contributed to the national economy the second largest sum in dollar earnings and the greatest number of employment positions of all agricultural industries, aerial spraying was the most "cost-effective" option for contracted small growers. Activists denounced this assertion vehemently, saying, "Cost effective for whom? Do we realize the real costs of the practice? So far the current computations of cost does [sic] not even include the cost to health, environment, and other organisms that are affected."[56]

Arguing that the ordinance was an "unreasonable exercise of police power" and that the ban was "tantamount to confiscation of property without due process of law,"[57] Davao Fruits Corporation and Lapanday Corporation, members of the Pilipino Banana Growers and Exporters Association, filed a civil lawsuit in the Davao City Regional Trial Court. Their legal filings questioned the constitutionality of the city-wide ordinance banning the sprays in April 2007.[58] While the district court upheld the ordinance in a decision five months later, MAAS's victory was brief, as it prompted the growers association to raise the case to the Court of Appeals in Cagayan de

Oro in November that same year. At a number of these hearings, Dagohoy was present to testify, and he recounted the experience, saying, "The question to us from the banana plantations was whether we could provide scientific evidence. 'What is written down?' they said, 'If you got sick, where are the medical records?'" The banana industry demanded that allegations of ban supporters be substantiated with objective fact, "not just psychological and emotional tactics," to cite one of its representatives.[59]

MAAS was confronting the fact that by appealing to their molecular sovereignty and referring to broad categories like "spray," "toxic rain," and "chemical showers," they simply could not provide the correlations and causations deemed necessary for legal or scientific argument. When MAAS members listed the changes they saw in their bodies and in the bodies of their children, the list grew infinitely long and increasingly severe. In our conversations, they linked everything—such as rashes, asthma, bloated stomachs, stunted growth, and enlarged sexual organs, as well as male sterility, paralysis, diabetes, cancer, and sudden death—to the invasive chemicals. The lack of apparent consistency in the symptoms sat uncomfortably with the kind of rationality on which modern chemical regulatory institutions are built. At the heart of chemical science, both in terms of its regulation and in terms of activism against it, is the push to establish a cause-and-effect relationship between exposure, on the one hand, and manifestations of harm, on the other. In legal cases, medical investigations, and scientific experiments on toxicity, the singularity of that cause is key.

Scholars of science and technology studies have questioned the epistemological paradigm within chemical science that portrays chemicals as discrete and isolated entities.[60] The current "chemical regime of living," borrowing from Michelle Murphy, involves scientific understandings of chemical relations that are informed less by efforts to heuristically capture the ecological world as it is, or by commitments to protect and steward lives and landscapes, and more by drives to promote the interests of industrialized production of which chemicals are an inextricable part.[61] The notion of chemical discreteness exemplifies how innovations in the chemical sciences may render synthetic molecular relations detectable, even investigable, but primarily in ways that elude corporate accountability. The rub is that such understandings in turn organize environmental regulation as well, resulting in purposeful blindness to the realities of cumulative, chronic, interactive, and low-level exposures.[62] It also results in regulatory paradigms that are regrettably reliant on damage-based research and risk assessment, especially in the case of multiple or mixed chemicals for which there is no available data.[63]

When activism operates within current regulatory paradigms, sometimes there can be successful litigation. A notable example is the case of banana plantations in Central America, where a substance used to kill nematode worms—called DBCP or Nemagon—was singled out as exposed workers traced the devastating links to male infertility in a decades-long battle that ran from the 1970s to the early 2000s.[64] In Costa Rica, afectados built a scientifically authoritative case, engaged public support, and eventually won widespread compensation, though in ultimately small amounts.[65] In many respects, however, the DBCP case is more the exception than the norm. Many researchers, such as environmental anthropologists and chemical geographers, have noted that victims of chemical fallout regularly find themselves in situations where demonstrating causal relationships is impossible due to spatial and temporal lags.[66]

The aerial spray case stayed in the Court of Appeals in Cagayan de Oro for over a year. At the close of 2008, MAAS activists camped on the sides of roads outside the courthouse for two-and-a-half months to push for a resolution. They also took their demonstrations to Manila, where they marched before the Department of Agriculture bearing pictures of their children and their homes on placards strewn around their necks. Wherever they went, they left cut up banana blossoms—in Visayan, "puso sang saging," "banana hearts"—strewn on the ground, a symbolic reference to the heartlessness of corporate greed. The court reached a decision in January 2009, reversing the Regional Trial Court's ruling and declaring the Davao City ban on aerial spraying as "unconstitutional." The ruling demonstrated that legal structures, formatted around particular evidentiary politics, were simply not built to meet the campaign's demand for molecular sovereignty. Neither were they designed to provide citizens the vision of human dignity they desired. MAAS elevated the case to the Supreme Court of the Philippines in October that same year, still unaware that the decision in Cagayan de Oro would mark the beginning of the Pilipino Banana Growers and Exporters Association's legal triumphs in the succeeding years.

Conclusion: "All We Have Are People"

On August 16, 2016, MAAS received crushing news. The Supreme Court of the Philippines had declared Davao City's ordinance banning aerial spraying as "invalid" and "unconstitutional."[67] The decision was based on three legal grounds, but here I mention only one. The court justices deemed the ban in violation of the Philippine Constitution's *equal protection clause*, which states that no person should be "denied the equal protection of the law":

> It can be noted that the imposition of the ban is too broad because the ordinance applies irrespective of the substance to be aerially applied and irrespective of the agricultural activity to be conducted. [...] The imposition of the ban against aerial spraying of substances other than fungicides and regardless of the agricultural activity being performed becomes unreasonable inasmuch as it patently bears no relation to the purported inconvenience, discomfort, health risks and environmental danger which the ordinance seeks to address.[68]

To summarize, the Supreme Court saw it unclear that the broad banning of an agricultural method, irrespective of its content—chemical or otherwise—would bear any relation to alleviating human and environmental damages. The decision, a nefarious warping of Davao residents' constitutional rights, came only a few months before I would arrive for fieldwork and begin piecing together the story of why the MAAS campaign unfolded in the way that it did. Until this day, the export banana sector is the only agricultural industry in the Philippines that employs the controversial method of aerial spraying. There are still no specific government laws regulating its use.

I visited Dagohoy to talk about the Supreme Court decision, and he opened our conversation on a curious note. "Do you know any artists?" He said he fantasized about commissioning a satirical depiction of the justices of the Supreme Court with animal heads on their shoulders (see figure 3.4). He was determined to get a draft he made on his computer into a national newspaper in time to make some waves for No Pesticides Day and the anniversary of the Bhopal Disaster on December 3rd.[69] Dagohoy's artwork, entitled "Supreme Court of Animals," never did find its way to print. However, the following comment appeared in the regional paper *SunStar Davao*: "Magaway said he perceived the justices of the SC as inhuman and no different from animals."[70] For the MAAS activist, there was a bitter sense of irony that the same government should recognize an aerial spray ban in other provinces like Bukidnon and Cotabato. There, provincial ordinances had been motivated by corporate demands to protect not exactly its residents but rather the livestock industries. This *political economy of speciation*, a term borrowed from Alex Blanchette, "reverses the typical hierarchy of species" and traps humans in worlds structured primarily to support the "fragile capitalist life forms" of factory-farmed animals.[71] "What is real justice," Dagohoy asked me, "when the laws are governed by the wind [Tagalog: batas ng hangin]?"

3.4 The Supreme Court of Animals in the Philippines, by Dagohoy Magaway, 2017. Photograph by Dagohoy Magaway.

This was a poignant allusion to the aerial sources of his despair. Feeling the full blow of the decision from the highest tribunal, the man was visibly filled with rage but also renewed conviction about his advocacy and vocation to the community. "I will not stop speaking... they cannot silence my humanity..." In a poetic move to reclaim that humanity in the hierarchy of animals, he continued to say, "I see the justices of the Supreme Court as water buffalo, cows, goats, and chicken. In the entire province in Bukidnon, they banned aerial spray in 2001. Why? Because there are piggeries and poultries, cattle ranches, and livestock! Because this is a business! They cannot spray because they would kill off the chicken. Is that how the justice system is going to be here in the Philippines, where animals are more favored than people? They have the hearts and the commitment of beasts [Tagalog: hayop na pakiramdam, hayop na paninindigan]"!

How fortunate that Bukidnon should have precious livestock, Dagohoy reflected, as it seemed that only commodifiable crops or animal lives were deemed worthy of government protection and corporate accountability. "Here in Davao, all we have are people."

Throughout the campaign, activists' disavowal of the subhuman has empowered victims in their pursuit of dignity and allowed them to navigate around the dead ends of chemical regulatory paradigms focused on identifiable and isolated chemicals. As I hope this chapter has made clear, their politics of exclusion, which envisioned human dignity at the expense of other forms of life, was an ethical choice that activists felt they had to make in the face of practical environmental and legal limitations. For many subjects who continue to live through the legacies of colonialism beyond the Philippine banana belt, there often seems to be no recourse but to insist on a delineation, rendered fundamental, of the human and the nonhuman, and on a hierarchy between the two. At the same time, for Dagohoy and his fellow activists at MAAS, this came with the stinging realization that *if only* human beings had been awarded the kind of value that commercial livestock had, then perhaps the Supreme Court would have ruled in favor of their protection.

Interspecies competitions for inclusion in the ambit of justice reoccur wherever multiple forms of oppression collide and differentially subjugated communities come face-to-face. Mainstream animal rights advocacy, as Bénédicte Boisseron reflects in her book *Afro-Dog*, frequently construes Black lives and animal lives as caught in a tug-of-war and a competition of value in the ways it uses the project of anti-racism as a platform for anti-speciesism.[72] Echoing dynamics seen in the MAAS campaign, Boisseron explains, "Such a state of affairs is bound to eventually create a counteraction where Blacks no longer want to be 'ranked together' with the animal and might even want to, in the 'scale of being,' *matter more* than the animal, what one may see as overcorrection or simply a preemptive defense against the weight of history."[73] Certainly, to speak of multispecies justice is to acknowledge this space of tension, where anthropocentric and more-than-human claims to integrity may sometimes be at odds. They are at odds because actors find themselves operating within compromising structures that seem to leave them little choice but to invest in principles of state bureaucracies that are contractual and possessive in nature.

Yet, it is vital to remember that even these forms of counteraction fail to acknowledge that human life comes to be subjugated by way of the subjugated animal. Human communities rendered as "pestiferous" are lives endangered because the animal pest is so unquestionably subject to extermination. "The specter of the animal still influences the fate of racial differences," according to Boisseron.[74] In a similar vein, Che Gossett suggests, "Blackness has already been racialized *through* animalization."[75] Thinking alongside these scholars but in the context of Mindanao's plantation zones thus compels the questions: What would environmental activism look like if we recognized these processes as truly combined? How do we move beyond the interspecies exclusionary project, and beyond the notion that animalization—embedded in the assertion "We Are Not Pests"—can only be humiliating, reductive, or trivializing? On Davao's plantation peripheries, we might look to the divergence already emerging between activists' rhetorical strategies and the material experience as an opening. While rejecting their relegation to the state of the animal or the object, locals simultaneously recognize that vulnerabilities are violently distributed across species divides. They understand the workings of interspecies transcorporeality in their turn to plant and animal sentinels as ways to see and understand the harm on their own bodies.[76]

Civic organizations know all too well that the true costs of aerial spraying include both the health of the human community as well as that of the environment and other organisms affected. These experiences reveal that forms of oppression are not only structurally shared but truly combined at the visceral and corporeal level.[77] Can we envision activist rhetoric that speaks this truth most faithfully, that defies the forces, however powerful, that compel actors to believe that there is no recourse but to insist on maintaining a hierarchy of value distinguishing the human from the nonhuman? Can we compose an alternate vision of justice that does not flatten the incommensurate violences chronicled above, but that also does not accede to dominant plantation logics, which can only construe the dignity of biological and ecological life at the expense of others? How can competitive justice be made coalitional?

In searching for sources of inspiration, we might heed the invitation of literary voices. As Joshua Bennett has illuminated by looking to African American traditions, "rather than triumphalist rhetoric that would eschew the nonhuman altogether, what we often find instead are authors who envision the Animal as a source of unfettered possibility."[78] When tools for derision are embraced, Bennett shows through the writings of Frederick Douglass, fraught proximities between the human slave and the nonhuman animal become "kinship forged in the midst of unthinkable violence, kinship born of mutual subjugation."[79] The figure of the pest, too long used as one such tool for denigration and extermination, is ripe for reclaiming. As Clapperton Chakanetsa Mavhunga has powerfully declared, "To call them pests is to acknowledge their resistance or that they simply moved about in ways the colonial regime saw as resistance."[80]

How would we articulate such an alternate vision of multispecies justice? I can think of no better words to cry out, and no better imaginary to labor toward, than "We, Too, Are Pests!"

Notes

1. Rose, *Reports from a Wild Country*; Rose, *Wild Dog Dreaming*.
2. Giraud, *What Comes after Entanglement?*
3. Giraud, *What Comes after Entanglement?*, 7. See also Shah, "Against Eco-Incarceration."
4. Chao, "In the Shadow of the Palm."
5. Celermajer et al., "Justice through a Multispecies Lens"; Cripps, "Saving the Polar Bear"; Schlosberg, "Ecological Justice for the Anthropocene."
6. Ticktin, "Invasive Others," xxv; Raffles, "Jews, Lice, and History."
7. Foucault, *History of Sexuality*; Foucault, *Society Must Be Defended*; Guthman and Brown, "Whose Life Counts"; Li, "To Make Live or Let Die?"
8. Manapat, *Some Are Smarter Than Others*, 145; Wiley, *The Banana*, 45.
9. David et al., *Transnational Corporations*, 26.
10. David et al., *Transnational Corporations*, 26.
11. ICL Research Team, *Human Cost of Bananas*, 63–65.
12. See also Chao, "The Beetle or the Bug?"
13. Shapiro, "Attuning to the Chemosphere."
14. Lowe, "Viral Clouds," 644.
15. National Research Council, *Animals as Sentinels*; Van der Schalie et al., "Animals as Sentinels."
16. Keck and Lakoff, "Sentinel Devices"; Keck, "Livestock Revolution."
17. Keck, "Sentinels for the Environment." For reflections on transcorporeality, see Alaimo, *Bodily Natures*.
18. Choy, *Ecologies of Comparison*; Choy, "Distribution."
19. Guthman and Brown, "Whose Life Counts."
20. Keck, *Avian Reservoirs*, 351.
21. Nading and Fisher, "*Zopilotes, Alacranes, y Hormigas*."
22. Shapiro and Kirksey, "Chemo-Ethnography," 489.
23. See also Chao, "The Beetle or the Bug?"
24. Tsing, Mathews, and Bubant, "Patchy Anthropocene"; Tsing et al., *Feral Atlas*; see also Warren, *Brazil and the Struggle for Rubber*; Giesen, "Herald of Prosperity"; Marquardt, "Green Havoc"; Soluri, "People, Plants, and Pathogens"; Soluri, "Accounting for Taste"; Rose, *Reports from a Wild Country*; Rose, *Wild Dog Dreaming*.
25. Besky, "Exhaustion and Endurance."
26. Clauses in grower cooperatives' contracts with large exporting companies often stipulate that while spraying schedules were under the full control of the latter, the expense was to be shouldered by the former. For example, the following line under the "Rights and Obligations of the Seller" appears in a contract between a Davao del Norte-based Agrarian Reform Beneficiaries Cooperative and a large banana exporter: "Permit the BUYER or its agent to carry out aerial Sigatoka control activities charged to the SELLER's account and expense at cost. If the BUYER, however, fails to provide Sigatoka control activities, the SELLER upon written notice to the BUYER has the option to engage the services of other aerial spraying

companies. The SELLER agrees to eliminate, at its own expense, trees and other obstructions that will interfere with aerial spraying." Similar language appears in several contracts that the author has reviewed.

27 Paredes, "Chemical Cocktails Defy Pathogens."
28 Cholinesterase is an enzyme that is central to the normal functioning of the nervous system. Pesticides that are considered as "cholinesterase inhibitors" disable this enzyme, resulting in symptoms of neurotoxicity, including tremors, nausea, and weakness at low doses and paralysis and death at higher doses.
29 Other Sigatoka-control fungicides that are not considered as "PAN Bad Actors" include difenoconazole (Sico), bitertanol (Baycor), tridemorph (Calixin, Tokalix), tebuconazole (Folicur), spiroxamine (Impulse), Pyrimethanil (Siganex), trifloxystrobin (Twist).
30 Murphy, "Chemical Regimes of Living"; Murphy, "Afterlife and Decolonial Chemical Relations"; see also Agard-Jones, "Bodies in the System."
31 Dumit, "Writing the Implosion."
32 Mendez et al., "Tracking Pesticide Fate in Conventional Banana Cultivation in Costa Rica"; Grant, Woudneh, and Ross, "Pesticides in Blood from Spectacled Caiman (Caiman Crocodilus)"; van Wendel de Joode et al., "Aerial Application of Mancozeb and Urinary Ethylene Thiourea (ETU) Concentrations."
33 Murphy, "Afterlife and Decolonial Chemical Relations," 495; see also Myers, *Rendering Life Molecular*.
34 Barad, *Meeting the Universe Halfway*; Fortun, *Advocacy after Bhopal*; Fortun, "Ethnography in Late Industrialism."
35 Brent and Hollomon, *Fungicide Resistance in Plant Management*; Fungicide Resistance Action Committee, "2016 Meeting Minutes."
36 Estremera, "I Am Not a Banana" (italics added).
37 Van den Bosch, *Pesticide Conspiracy*. See also Galt, *Food Systems in an Unequal World*; O'Brien, *Making Better Environmental Decisions*.
38 Langwick, "Politics of Habitability."
39 Fanon, *Wretched of the Earth*; Mbembe, *On the Postcolony*; Weheliye, *Habeas Viscus*; TallBear, "Indigenous Reflection on Working."
40 See also Rice, *Dean Worcester's Fantasy Islands*.
41 Dacudao, "Abaca."
42 Tompkins, *Racial Indigestion*.
43 Fanon, *Black Skin, White Masks*.
44 Mavhunga, "Vermin Beings."
45 Anderson, "Politics of Pests."
46 Ticktin, "Invasive Pathogens?"
47 Ticktin, "Invasive Pathogens?"
48 Comaroff, "Invasive Aliens"; Kim, "Invasive Others and Significant Others."
49 Haraway, *When Species Meet*.
50 Agard-Jones, "Ep. #35 – Vanessa Agard-Jones."
51 This included Balingain, Tugbok, Sirib, Wangan, Tamayong, Lacson, Tigato, Dakudao, and Manuel Guianga.
52 "Yap Intervenes in Aerial Spray Ban."

53 Nikol and Jansen, "The Politics of Counter-Expertise on Aerial Spraying"; "A Timeline of Selected Documents." Nicol and Jansen, *The Philippine Controversy over Aerial Spraying of Pesticides.*
54 "Aerial Spraying 'Safest, Most Effective.'"
55 Arquiza, "Davao City Govt, Farmers Push Ban on Aerial Pesticide."
56 Bantay Kinaiyahan, "People vs. Profit."
57 Arquiza, "Davao City Govt, Farmers Push Ban on Aerial Pesticide."
58 Nikol and Jansen, "Politics of Counter-Expertise on Aerial Spraying."
59 Balane, "Banana Firms."
60 Boudia and Jas, *Powerless Science?*
61 Murphy, "Chemical Regimes of Living."
62 Boudia and Jas, *Powerless Science?*
63 Boudia and Jas, *Powerless Science?*
64 Bohme, *Toxic Injustice.*
65 Bohme, *Toxic Injustice.*
66 Guthman, "Lives Versus Livelihoods?"; Guthman and Brown, "Whose Life Counts"; Harrison, *Pesticide Drift*; Romero et al., "Chemical Geographies."
67 Nonato, "SC Voids Ordinance vs. Aerial Spraying."
68 "Mosqueda v Pilipino Banana Growers and Exporters Association" 2016.
69 Fortun, *Advocacy after Bhopal.*
70 Tejano, "Anti-Aerial Spray Group Hopes SC Decision to Change."
71 Blanchette, "Herding Species," 641; Blanchette, *Porkopolis.*
72 Boisseron, *Afro-Dog*; for canonical writings in Western animal rights advocacy, see Singer, *Animal Liberation*; Spiegel, *Dreaded Comparison*; Patterson, *Eternal Treblinka.*
73 Boisseron, *Afro-Dog*, 34.
74 Boisseron, *Afro-Dog*, 8.
75 Gossett, "Blackness, Animality, and the Unsovereign" (italics added).
76 Alaimo, *Bodily Natures.*
77 For more reflections on viscerality, see Blanchette, *Porkopolis.*
78 Bennett, *Being Property Once Myself*, 3.
79 Bennett, *Being Property Once Myself*, 1–2.
80 Mavhunga, "Vermin Beings," 156.

4

Prison Gardens and Growing Abolition

Elizabeth Lara

4.1 (*previous page*) Original drawing by Feifei Zhou.

WHEN RUTH WILSON GILMORE DESCRIBED "breathtakingly cruel twists in the meaning and practice of justice" in her canonical text *Golden Gulag*, she was referring specifically to two infamous laws added to the late-twentieth-century California penal code. These laws had been central in helping to fill (and quickly exceed the capacity of) the numerous state prisons that, at the time of her writing in the early 2000s, were either recently built, being built, or slated for construction.[1] With the passage of time, such cruel and distorted theories and practices of justice continue to underlie virtually every aspect of the contemporary carceral state. Beyond the profusion of actual prison structures throughout much of the world, carceral logics and technologies— apparatuses of punishment, surveillance, immobilization, and coercion— have long been developed within other physical and figural spaces.[2] Often by design, this process generates carceral geographies, expansive regimes that skew conceptions of justice and reorient our bodies and our modes of knowing and being in the world.[3] Simultaneously, *abolition* geographies are solidifying through distinct forms of place-making work—liberatory gestures, sites, and relationships that are being renamed and reimagined within a gradually cohering project.[4] Gilmore describes abolition geography as something "we have to make from the carceral geography [in which] we live."[5] Communities build abolition geographies within or proximal to the carceral state in ways that remove or divert its power and resources, or prevent its accumulation in the first place.

In summer 2020, impacts of the coronavirus pandemic coincided with the incessant legacies of anti-Black violence in US policing and politics. As unprecedented numbers of people mobilized worldwide in defense of Black lives, the distinction between reform and abolition (particularly in relation to policing institutions) was widely discussed within social movements and popular media outlets. Discourse on abolition emerged in a way not seen since the end of chattel slavery in the nineteenth century.[6] A growing number of people began taking note of long traditions of abolitionist modes of thinking, relating, and becoming.

Abolition as Multispecies Justice

Histories of racial capitalism are histories of all capitalism and are comprised of legacies of violent dispossession, colonialism, and slavery that originate from intra-European social structures and now permeate the world. Racial capitalism is a system through which racialization is an ongoing process comprised of hierarchy making through differentiation, induced from structures of economic and military power.[7] These processes have created the preconditions for and continue to deepen the many socioecological crises that define the present day. Inextricable from the ongoing subjugation of racialized populations, the lives and deaths of plants and animals, similarly deracinated and rendered fungible, have been central to the emergence, maintenance, and undoing of innumerable carceral regimes throughout history. I therefore claim that *the work of abolition is also the work of multispecies justice.*

Multispecies justice is not a carceral justice. It does not come through expanding or reforming the purview of the current dysfunctional and illegitimate arbiters of the criminal legal system.[8] It is an intersectional justice that comes through abolition. Abolition is not merely a matter of dismantling prisons as exclusively death-making institutions.[9] It is also acutely concerned with responding to the needs of those most impacted by policing and incarceration through designing and building socioecological relationships and systems of support which make prison and its adjacent institutions unnecessary.[10] My hope is that proposing abolition and multispecies justice as deeply coconstitutive does not have the effect of collapsing matters of power, race, class, gender, and other differences that are sometimes obscured by multispecies frameworks.[11] Rather, I extend this proposition as one way to hold liberation central in narratives, theories, and frameworks that inevitably oscillate between gestures that are flattening and those more attentive to critical differences.[12]

I offer this articulation in solidarity with the growing number of scholars, artists, organizers, healers, and many others who are investing in the pluralist practices of visualizing abolition.[13] Such practices are necessary given that an ultimate abolitionist future does not, perhaps cannot, exist. The same can be said for aspirations of justice.[14] Pursuits of abolition and justice alike are not hindered by concerns of personal risk or unclear end goals; they are mobilized by the urgent demands of the present, by "a love without future, a kind of fullness of being."[15] This work, therefore, is motivated less by a clear and coherent set of evidence and more by an intuitive drive to invite exchange between abolitionist traditions and burgeoning theories of multispecies justice. I offer this work as a potential bridge, as grounds for further intersectional coalition building, and as a mode of reiterating what many elders in the struggle have long understood—that it is not only human persons who stand to benefit from the realization of a world without prisons.

Bridging abolition and multispecies justice requires a series of reckonings with the many valences contained in conceptions of the nonhuman as they move through the carceral landscape. These reckonings include, foremost, continuing to confront the ways in which white supremacist, heteropatriarchal, ableist notions of humanity are fortified to otherize numerous groups of people who are dehumanized by systems of state violence, policing, and incarceration.[16] In building bridges between abolition and multispecies justice, it is also important to reckon with the damaging erasures and simplifications regarding precisely who is ensnared and trying to survive within prison infrastructures and carceral ecologies.

Stereotypical images cast prisons as lifeless, except for the human beings contained within. While the surfaces of modern facilities appear stark, sterile, and highly disciplined, ecological assemblages cocreate past and present prison worlds in many ways. In twenty-first-century US prisons, a vast range of species (and innumerable individuals) are mobilized and instrumentalized in the name of rehabilitation, therapy, environmental education, food security, sustainability, job training, and public relations.[17] Of the many multispecies carceral modalities, a few examples that have received scholarly attention include the following: the kitchen gardens of enslavers and enslaved people; the mobilization of plants and animals considered capable of assisting in projects of "prisoner rehabilitation"; convict leasing and prison labor in agricultural production; as well as the countless species deemed companions, weeds, pests, scavengers, or contraband while inhabiting carceral facilities.[18]

In California, a prison might meet a state mandate by diverting a portion of its organic waste to an industrial composting facility, and then use compost from that same facility for topsoil remediation prior to a garden installation. Such developments, however, cannot be read simply in terms of an administration that is increasingly imagining its prison as part of a local ecology (i.e., becoming "green"), when in fact prisons are always already generating ecologies.[19] Legal advocates, journalists, and others have documented how carceral facilities are often built on and actively producing toxic landscapes.[20] There are enduring correlations between environmental pollution and prison construction and operation.[21] Long-term proximity to US prisons often results in illness and premature death within populations of prison staff, incarcerated people, and local communities.[22]

No matter how expansive, an ecological consciousness and imagination delimited by the logics of the prison-industrial complex remains damaging.[23] Although anthropocentric projects of multispecies manipulation are constitutive elements of prison socioecologies, animals, plants, microbes, and viruses have always enacted their own oppositional agendas. Further, liberatory action and theorization continue surging within the heart of the carceral state.[24]

Interspecies Accomplices

Of the many avoidable disasters to emerge during the coronavirus pandemic, one of the most devastating and unrelenting was the failed state response to the impacts of COVID-19 on incarcerated communities and their loved ones.[25] In California, more than 113,000 additional cases of the disease recorded during summer 2020 have been attributed to policy and practices involving the state's carceral facilities.[26] Data gathered from across the country showed that some of the worst outbreaks and largest clusters of COVID-19 occurred in densely populated spaces with varying degrees of forced confinement, where physical distancing was often impossible. While the overwhelming majority of such carceral clusters formed in prisons, jails, migrant detention centers, and universities, others occurred within food processing plants, factories, elder care facilities, psychiatric hospitals, and ships.[27]

In prisons and jails in particular, the provision of cleaning supplies and personal protective equipment to incarcerated people was inadequate, if not intentionally neglected. Some departments of corrections incarcerate high rates of elderly and medically vulnerable people, and many lack adequate healthcare services. Therefore, before the effects of the pandemic even took hold in the United States, medical experts and public health employees expressed extreme concern regarding the well-being of incarcerated people. Across the country, the COVID-19 death rate among the imprisoned population eventually peaked at a rate three times higher than the nonincarcerated public.[28]

Ruth Wilson Gilmore defines racism as "the state-sanctioned or extralegal production and exploitation of group-differentiated vulnerability to premature death."[29] As Black, Latin American, and Indigenous people within the United States have been disproportionately incarcerated *and* disproportionately impacted by COVID-19 due to centuries of structural inequity, it is increasingly apparent that vulnerability to premature death continues to occur by design.[30] One way that Gilmore's definition of racism is useful and actionable is that it encourages thinking systematically about relationships and the design of things, "so that it's possible to see how various groups come into being and therefore then become naturally available for organizing and struggle."[31] Bénédicte Boisseron's work similarly expands notions of who is available for organizing and struggle, as she "reject[s] the instrumentalization of the comparison between racialized human beings and animals." Boisseron focuses on "interspecies alliances," confronting "entangled forms of oppression" in order to "defy the construction of blacks and animals as *exclusively* connected through their comparable state of subjection and humiliation."[32] While the coronavirus pandemic amplified consequences of structural racism and other long-standing crises, it also created openings through which old power dynamics might be disrupted by coalitions among people and other beings involved in multispecies projects of co-becoming, healthfulness, and healing.[33]

Reckoning with the sedimented impacts of systemic racism and inequality, amplified by the pain and trauma of the pandemic, I cannot help but read this historic moment—this prolonged catastrophe, this intensification of the catastrophe that prisons have always been—through the life-affirming work happening in certain gardens. Throughout the years that I have spent gardening throughout California in places like prisons, carceral heritage sites, and a militarized border wall, I have often felt the need to resist lingering too long in the figurative realms that accompany such sites.[34] These are the realms where soil quality is a proxy for human culture and roots are likened to the many forms of vital work that go unnoticed. Seeds are ideas with limitless potential. Upturned weeds and sifted compost are likened to the removal and regeneration of negativity and death into something promising. Companion planting (putting different species together for symbiotic benefits) provides lessons on the power of mutual aid and diversity of tactics. The potency of such metaphors increases in sites of direct conflict and turmoil, where gardens begin to resonate with meanings and gestures that point to the fundamental resilience of all life.

Such figurations, I often felt, were secondary to the embodied, visceral qualities of the gardening work itself. I tended to dwell more on the sensuous immediacy of a big group of people building a garden where there was no garden before, or of the whizzing hummingbird eagerly dipping down toward a newly arrived plant that is yet to even be transplanted from its pot, or of the hard thwack that reverberates through my core when a pickaxe breaks into the sunbaked, clay-rich, highly compacted earth of the prison ground. But in March 2020, when the impacts of the coronavirus pandemic reached California, the numerous gardens where I had been working became inaccessible. Confrontation with a novel form of enforced distance left me more inclined to embrace what little of those spaces did remain accessible—not the gardens' actual lavender labyrinth or footpath inspired by a stellar constellation but their less tangible visionary material and their continued capacity to mobilize action and cohere relationships. In his book *Gardens: An Essay on the Human Condition*, Robert Pogue Harrison asserts that "gardens after all are never either merely literal or figurative but always both one and the other."[35] Perhaps then there is something literal about the way some prison gardens grow beyond the boundaries of their prisons.

Throughout space and time, manifestations of prison gardens vary widely. While my time spent in gardens at state prisons was facilitated by a volunteer program, gardens that emerge through formalized structures tend to be the exception. Throughout the long histories of human confinement, informal or illicit gardens have taken a far broader range of forms, emerging as gardens tend to do among virtually all sedentary populations with any access to seeds, soil, sun, and water. In some places, incarcerated people receive only staple food provisions from prison administrators, which they are obligated to supplement through outside support and access to small garden plots. For the incarcerated herbalist who tends medicinal plants otherwise deemed weeds, sometimes the garden is a regular prison yard.[36]

Despite their varied formations, there are qualities held in common among many prison gardens. They are sites of sustenance, and of refuge and repose, where humans and plants are coconspiring—both acting and literally breathing together toward shared futures, transcending the facts of their mutual physical confinement.[37] Timothy Choy writes of conspiracy in terms of the distribution of conditions that constrict and create the capacity for breath in late industrialism.[38] Prison gardens are places for practicing the mutually enriching qualities of gaseous exchange and appreciating the oxygen-producing, breath- and life-enabling qualities of photosynthesis.

These places may be enclaves, set apart from the surrounding conditions that make it hard to breathe. Early in the coronavirus pandemic, Achille Mbembe wrote of "the universal right to breathe." The realization of this right requires war—"if war there must be"—against not just a virus but against "everything that condemns the majority of humankind to a premature cessation of breathing, everything that fundamentally attacks the respiratory tract, everything that, in the long reign of capitalism, has constrained entire segments of the world population, entire races, to a difficult, panting breath and life of oppression."[39] The universal right to breathe, Mbembe explains, "is *an originary right to living* on Earth, a right that belongs to the universal community of earthly inhabitants, human and other."[40] This is a right similarly asserted through an abolitionist multispecies justice.

Of the gap between the act of *speaking about* justice and the process of *working for* justice, poet Ross Gay says that "tending to what one loves feels like the crux." Perhaps unsurprisingly, his curiosity "about a notion of justice that is in the process of exalting what it loves" is something he uncovered in part through his experiences as a devoted gardener and community orchard manager.[41] Prison gardens then, where exaltation occurs among certain gardeners, plants, birds, insects, and other lifeforms, contain possibilities for and processes of multispecies justice making. Situated at the interfaces of carceral and abolition geographies, they continue manifesting themselves as loving allies and accomplices, even (or especially) during the pandemic-induced lockdowns, rendering borders and boundaries porous, sustaining inspiration for radical thought and action.

Gardens Lost to Lockdown

Beth Waitkus founded Insight Garden Program at San Quentin State Prison in 2002 as a way of becoming capable of response in a world still reeling from September 11, 2001, and the United States' invasion of Afghanistan (see figure 4.2).[42] Given the broad range of documented psychosocial benefits offered to people by gardens, there was nothing particularly surprising about the changes Beth noticed as she and the first group of incarcerated participants established a garden on a medium-security prison yard.[43] Beth observed the way the garden attracted attention and use. She noted that it offered refuge, as gardeners and prison staff expressed feelings of reduced stress. She also saw that the garden was often one of the few racially integrated parts of the prison yard, as long-standing tensions were momentarily replaced by feelings of collective joy, accomplishment, pride, and hope.[44]

4.2 The garden at San Quentin State Prison, H-Unit. Photograph courtesy of Insight Garden Program.

Nearly twenty years later, on the other side of a pandemic that traumatized incarcerated communities worldwide, the program persisted. Relative to other prison garden programs, and to the status quo of prison programming generally, Insight Garden Program was unique in its duration, scale, and operational ethos. After its initial success at San Quentin, the program began developing a year-long curriculum as it eventually expanded into ten men's, women's, and youth prison facilities across California. Insight Garden Program was also unique in that the staff members were not prison employees but instead part of a nonprofit, community-based organization.[45] Furthermore, the program sustained vital inside-outside solidarities through a thriving volunteer base, providing rare opportunities for diverse groups of people to spend time in spaces of deep discussion and colearning.

Through the persistence of Insight Garden Program, incarcerated gardeners and their gardens disrupted power dynamics and dislodged prison logics. Subtle forms of disruption were written into the program's basic functions, in some cases starting before the garden itself was even built. Program participants held central roles throughout the processes of garden design, creation, and maintenance—work that proceeded through at times highly precarious circumstances involving continual negotiations of agency and access. Negotiations occurred at all levels: from the administrative affordances that made state-salaried correctional officers available to unlock the gates that led to the gardens; to the sometimes astonishingly slow shifts in institutional culture that brought formal permission for gardeners and program managers to actually eat what they grow; to the ultrasonic animal repellants that were installed as a nonlethal way to ward off hungry ground squirrels that straddled the line between companion and pest. Through acts of literal unearthing, these multisensory spaces ruptured the fabric of the prison, reorienting bodies away from the vast carceral horizon, toward more intimate microclimates and microlandscapes.[46]

These spaces were resignified in part through what Sarah L. Lincoln calls "fugitive gardening" or "a radical appropriation of hegemonic spaces and practices that [...] deconstructs the logics of mastery and hygienic possessiveness that underpin colonial culture."[47] Tending to gardens was one way for incarcerated people to do the world-making work of influencing how prisons fit (or fail to fit) into our physical, mental, and spiritual worlds.[48] Crucially, it was not merely the gardens themselves but also the relational networks through which the gardens were activated that produced openings toward multispecies justice and enabled anticarceral dreaming and doing. Groups of participants engaged in the creative and collective labors of caring for soils, plants, self, and one another. They did so in ways that flowed against the powerful influences that induce vulnerabilities among all who live long term in environments plagued by poor air and water quality.[49]

Coherence arises between contemporary prison gardens and gardens in other sites or eras defined by oppression and violence. Janae Davis and her colleagues, for instance, write that the garden plot for enslaved people "might help conceptualize multispecies assemblages that lead out of socioecological crises toward better futures—assemblages that are not just envisioned but lived and that simultaneously tend to the needs of social reproduction, social justice, and ecological care."[50] The assemblages of justice and care tended through the garden program's operations functioned to offset the material toxicities that define prison life while simultaneously counteracting the cultural toxicities that run rampant in such institutions.

However, there is a dualism within such gardens that must remain unreconciled.[51] Prison gardens that are tended in the name of "personal transformation" and "rehabilitation" arguably produce a "better" institution. At the same time, they also undermine standard prison operations that function to isolate people who are often defined foremost by the criminal acts for which they have been sentenced. As Insight Garden Program staff and volunteers did critically important work to counter the causes and impacts of incarceration, they simultaneously fostered close partnerships with prison administrators. This dynamic may ultimately lend violent institutions a degree of legitimacy and palatability while potentially obscuring broader, underlying concerns. Michelle Brown thus describes the prison garden as "the garden of reality," where "cultivation requires a complex violence—a recognition of the wasteland and a commitment to something else in that very place, knowing that there shall otherwise be . . . no other way out."[52]

At an Insight Garden Program staff meeting in early March 2020, the anxiety was palpable as people named their various concerns about the encroaching pandemic. Some were worried about the elderly and medically vulnerable people they were working with, while others reflected on their prior experiences of prison facilities on lockdown due to other infectious disease outbreaks. A few attendees were frustrated about the fact that hand sanitizer was inaccessible to incarcerated people and considered contraband due to its alcohol content.[53] Everyone in the room was familiar with the precautions already commonplace among those working with the garden program—handwashing upon entering and exiting prison was habitual, and elbow bumps instead of handshakes (the only form of physical contact typically allowed) were standard practice, especially in facilities that had known cases of the highly contagious norovirus that causes gastroenteritis. A couple days after the garden program staff meeting, the California prison administration issued their first formal response to the pandemic by canceling all visitations. A few days later, they issued a statement declaring: "No volunteers or rehabilitative program providers allowed to enter prisons."[54] By early April, all facilities began functioning under the first of many "modified programs," requiring that incarcerated people be increasingly confined to their assigned living spaces.

With the onset of the pandemic, long-term relationships between program managers and their participant groups were suddenly altered. Relationships between many incarcerated gardeners and their gardens were severed. This change necessitated drastic pivots in priorities and activities and tested the commitment of the various organizations providing in-prison programs. For decades community-based organizations have been assuming the caretaking roles neglected by a state that was never designed to meet the needs of vulnerable, marginalized people. Long before the pandemic, these organizations have been taking on reparative, liberatory care work with and for incarcerated and formerly incarcerated people and their loved ones. Many networks and organizations were therefore well positioned to respond swiftly to the shifting terrain of the pandemic by forming new solidarities and repurposing preexisting interinstitutional infrastructure. As their standard operations were derailed, Insight Garden Program staff quickly activated their relationships with prison employees and other organizations to acquire and deliver thousands of masks and bars of soap to various facilities. They started focusing their operations on providing urgent, unanticipated forms of support for incarcerated people.

As the prison administration began to gradually make announcements about early releases of people from prison due to the heightened concerns caused by overcrowding, there was a logistical scramble. People were being abruptly released into pandemic-stricken towns and cities with no housing or transportation, and there was no single point of contact within the department of corrections for people trying to coordinate support services. Due in large part to the advocacy of family members and community-based organizations, the process quickly improved. For years, Insight Garden Program staff members have prioritized inside-outside continuity in their relationships with former participants who have left prison. The dissolution of in-prison programming only increased their emphasis on coordinating gate pickups and first meals, as well as on helping people access employment and housing opportunities, identification cards, health care, laundry, and other practical services. Along with other community-based organizations and volunteers, they started to provide essential supplies such as hygiene kits, masks, gift cards, food, and more. As the stress of the pandemic mounted, Insight Garden Program staff—a group increasingly comprised of formerly incarcerated people—also began inviting recently released people to virtual support-group meetings, hikes, and picnics.

Due to standard prison regulations against gestures that may indicate "overfamiliarity" between incarcerated and nonincarcerated people, community-based organizations were unclear as to whether the administration would permit them to conduct their programs via written correspondence with participants. Given the urgency of the situation and finding that the process of receiving formal permission was slow and decentralized, many organizations opted not to wait. Insight Garden Program participants received their first letters from the program staff around the same time the earliest official lockdown went into place. Opening with sincere greetings and an acknowledgement of the extreme changes brought by the pandemic, the letters also contained reflections on processing fear, anxiety, sadness, and other emotions. The text reassured people that the program would continue offering its services for those who would soon be released from prison. Any-

> Thank you for your letter of support as it was well received and vital during these difficult times. I felt that we would hear from the IGP team soon with information. Thank you. I will say, all is slow here as you know because of the outbreak of the virus which so far has not hit this place yet. I'm thankful for that.
>
> I do see and talk with few IGP inmates (███████, etc.) as we discuss the garden. I like to inform you all that although we're unable to visit the garden, however, I get as close to the fence as possible to hold conversation with the garden but of course she doesn't respond which let me know I've not gone "MAD" or something, Ha, Ha!
>
> The garden, she know we still love and misses her very much. I am unable to tell what is actually happening in those individual bonds [*binds*]. But whatever is going on, I like to think that her dirt and insects' home remains in tact. Which means that our ecosystem family continue to thrive in ways we've yet to discover.

4.3 Excerpt from a letter sent by an incarcerated gardener who had been unable to access the garden since the onset of coronavirus-induced lockdown.

one with an upcoming release date was encouraged to complete and return a "transition questionnaire," so that Insight Garden Program would be better prepared to assist them. The packet also contained a series of journaling prompts and a section on various meditation practices, including an invitation for participants to join the program staff in the practice of pausing at ten o'clock each Tuesday and Friday morning to meditate and envision the plants, insects, birds, sun, soil, and scents of the gardens they had cared for together.

After receiving the first packet in the mail, Ben, an incarcerated elder who had been participating in the garden program for about a year prior to the pandemic, was quick to write back. In his typewritten response he thanked the team for the contents of the letter, expressing his faith that, despite the cancellation of their weekly meetings, the now dispersed group of gardeners would soon be hearing from the program staff. Ben offered updates on how he remained in conversation with his fellow gardeners. Despite no longer being able to access it, he also remained in conversation with the garden, he wrote, getting "as close to the fence as possible" while out on the yard (see figure 4.3). I first read Ben's letter shortly after it arrived at Insight Garden Program's office. I have reread it many times since. While efforts to enact care despite the cruelty of enforced distance took countless heart wrenching forms during the pandemic, the tension Ben refers to—of finding who or what you love to be just out of reach—is always inherent to carceral social worlds. Ben's words situate his readers in the space of duality where prison gardens, and people's relations to and within them, are meaningful in ways that are simultaneously literal and metaphorical. And yet his words refer to something tangible and specific—the indefinite inaccessibility of a garden that he had tended and loved for many months. Since sending his first letter, Ben was transferred to another facility. While he was unfortunately unable to visit the garden a final time before he left, he continued to receive Insight Garden Program's correspondence and to respond with updates about his health and his eventual release after twenty-three years of incarceration.

No Public Health without Abolition

The pandemic shed new light on the ways that carceral modes of organizing life, labor, and justice have made consequential affordances to viruses and microbes that flourish in human bodies.[55] The first person in the custody of the California Department of Corrections and Rehabilitation reportedly tested positive for COVID-19 on March 22, 2020. At the time of writing, nearly fifty thousand incarcerated people and close to seventeen thousand staff members had become confirmed carriers of coronavirus. Official reports indicated that 255 of these people had died, including twenty-eight staff deaths.[56] The first major blunder occurred when San Quentin—a facility previously understood to be free of coronavirus—received a transfer of 122 medically vulnerable people from the California Institution for Men, then the site of the worst outbreak in the prison system. While the transferees had apparently tested negative for COVID-19, investigations showed that up to three weeks had passed since some of them had received their test. Several people were already showing symptoms of the disease during the hours-long bus ride north.[57] Those who were not yet symptomatic on arrival at San Quentin were housed on the uppermost tier of one of the prison's crowded, poorly ventilated units.[58] A state assembly member at a Senate Public Safety Committee meeting described the series of events that led to the outbreak as "the worst prison health screw-up in state history."[59] Still, less sensational "screw-ups" occurred far more regularly as transfers continued, and prison staff themselves transmitted the virus from facility to facility, yard to yard, dorm to dorm, and, inevitably, into their families and communities.[60]

Throughout the history of the prison institution, the relationship between care and punishment for incarcerated people has been troubling and difficult to distinguish.[61] During the pandemic, advocates for swift decarceration repeatedly argued the case that *prisons only know how to be prisons.*

Protection from exposure often meant being confined to a cell for most of the day. A positive COVID-19 diagnosis often resulted in being moved to solitary confinement. At the height of the San Quentin outbreak, Adnan Khan, executive director of the policy advocacy organization Re: Store Justice, tweeted about the conditions at a hospital that had received some of the more severe COVID-19 patients transferred from the prison. According to Khan's tweet, a hospital chaplain was told by prison employees that since incarcerated patients' bodies "belong to the state of California," the hospital did not have jurisdiction to contact and inform families of their patients' conditions. They could do so only if the patients died.[62]

While the impacts of the coronavirus reemphasized the systemic brutality and utter untenability of the prison system, they also created multiscale openings for change. The adult prison facilities in California have been notoriously overcrowded for years. The population of the prison system peaked in 2007 at over 200 percent of its design capacity, or more than 173,000 people.[63] While demands and mandates for population decreases therefore have a long precedent, the onset of the pandemic vastly increased public outcry and action regarding decarceration. Attorneys focused on litigation from precedent-forming cases linking overcrowding to the dismal quality of medical care, mental health care, and disability accommodations in state prisons.[64] The state senate held committee meetings to address concerns related to COVID-19 in prisons and the continual transfer of people from state custody to federal Immigrations and Customs Enforcement detention facilities. As virtual events became the norm, various organizations hosted countless online gatherings: strategy meetings, webinars, and urgent townhalls, as well as virtual ritual spaces for candlelight vigils and mourning the loss of those who had died from COVID-19 while incarcerated. For months there was a flurry of petitions, campaigns, phone zaps, Twitter storms, and the circulation of hashtags like #LetThemGo, #FreeThemAllForPublicHealth, and #StopSanQuentinOutbreak. There were also numerous in-person demonstrations. Protestors chained themselves to the gate of the state governor's home to demand mass releases, and numerous car caravan protests made use of horn honking and bold signage displayed from car windows to relay messages such as "Incarcerated Lives Matter" and "No State Execution By COVID-19!"

In addition to the huge loss of community and connection that resulted from the prison lockdowns, there was also a partial breakdown of the system of so-called time credits that incarcerated people in California rely on to potentially decrease their sentences, in part through participation in approved programs such as Insight Garden Program. As facilities went into lockdown, prison administration was not prepared to accommodate the shift from in-prison programming to correspondence-based programming, as they lacked protocol for how to administer and track credits for remote program participation. Although the time-credit system was instituted to facilitate much needed population decrease, the failure of administrators to meet mandated population reductions (to 137.5 percent of design capacity systemwide) was so persistent that even after decades of litigation, it was not until months into the pandemic that the overall prison population in California finally fell below this mark.[65] Amid these rapid changes, state officials announced and began to move forward with plans to close two prisons within two years, as fiscal and policy advisors and other advocacy groups began to make the case for up to ten closures.[66]

Prison programming and time credits, as well as gubernatorial sentence commutations, pardons, and pandemic-induced early releases, are all critically important tools for decarceration. However, the emphasis on these mechanisms for decreasing the prison population is misleading. They favor piecemeal, individualized quasi solutions while ignoring many of the broader systemic issues that led to rapid prison construction and so-called mass incarceration in the first place. The time-credit system is uniquely problematic in that it collapses concerns of rehabilitative programming and overcrowding, putting the onus on incarcerated individuals to remedy something that is ultimately a state policy-level concern. This moment in carceral history has a fraught inheritance—detailed expertly by many radical scholars of the criminal legal system—of deindustrialization, disinvestment, and organized abandonment; early- to mid-twentieth-century liberal efforts to end racial violence and bias in policing; the expansion of a legal system that criminalizes an increasing array of behaviors; the establishment of laws that lead to harsher prison sentences; neoliberal reforms aimed at economic efficiency; and other factors that have all led to what is more accurately described not as mass incarceration but rather the hyperincarceration of specific demographics, namely, people of color and poor people.[67]

This litany of concerns points to only a fraction of the sociopolitical context in which organizations like Insight Garden Program operate. It leaves virtually no room for imagining that a non-profit garden program might in any way help to undo the most bloated carceral system in human history. Further, such a program may begin to appear trivial when contextualized within the broad expanse of anti-carceral, abolitionist work that communities are undertaking. And yet, the practical strategies produced through decades of abolitionist organizing show the importance of growing ever-widening, multispecies systems of care.

"Something That Grows"

The work of abolition is also the work of multispecies justice, and it is something that grows alongside the countless pathways we make and find on our way through the physical and figurative gardens we inhabit. While these paths may lead through an at times seemingly barren or horizonless expanse, they demand that we keep moving, so that we too might, in the wise words of Ben, "continue to thrive in ways we've yet to discover" (see figure 4.3). Echoing Ben's words, the opening lines of Alexis Pauline Gumbs's essay "Freedom Seeds" provide a generative framing for understanding abolitionist organizing, shifting the emphasis away from absence and toward presence.[68] Gumbs offers garden imagery to recount lessons learned through the creation of an actual garden by her community. This garden emerged in the aftermath of a publicized rape trial that left many survivors, sex workers, women of color, and others vulnerable and retraumatized.

Her speculative opening line—"What if abolition isn't a shattering thing, not a crashing thing, not a wrecking ball event?"—invites her readers to orient themselves beyond simplistic understandings of abolition as a one-off event involving nothing more than the abrupt release of incarcerated people and the demolition of prison buildings. Gumbs continues, "What if abolition is something that sprouts out of the wet places in our eyes, the broken places in our skin, the waiting places in our palms, the tremble holding in my mouth when I turn to you?" Here, Gumbs references the inevitable pain and discomfort that occurs when community members decide to invest in responses to harm other than the dehumanizing, ineffectual options available through the police and the criminal legal system. A disclaimer abolitionists sometimes make when outlining the fundaments of their stance is that no one arrives to such a political position out of ignorance of the intractability of violence.[69] Rather, intimate familiarity with harm and its impacts is often precisely what drives people toward abolition and a recognition of the shortcomings of binaries like guilty/innocent and victim/perpetrator. Through Gumbs's imagery, we discover that on the other side of tearfulness, brokenness, vulnerability, and fear, there is something else. Conjuring images of nascent worlds slowly flourishing and becoming full so that they might be left still growing as old systems fall away, Gumbs prompts—"What if abolition is something that grows?"[70]

With these provocations, Gumbs harkens to one of the early lessons of the contemporary abolitionist project, namely, that it is insufficient to, in the words of Gilmore, attempt "to speak [prison] into illegitimacy" solely by highlighting its many flaws.[71] Similarly, Ruha Benjamin reminds us that "the facts, alone, will not save us" and that, to catalyze social change, the task at hand is to construct fictions, not as falsehoods, but as "refashionings" that "are not meant to convince others of *what is*, but to expand our own visions of what is *possible*."[72] During the pandemic, California prison gardens were virtually abandoned in the name of public health—and yet subtle gestures of care kept life in the gardens possible. At one facility, Insight Garden Program participants watered the garden with plastic bags filled from a drinking fountain. Elsewhere, incarcerated gardeners observed from a distance as weeds throughout their garden grew above their heads. Eventually, after a period of negotiation between program staff and prison employees, the gardeners were briefly granted access to tend their garden for a period of a few hours.[73] At another prison, one well-established garden became unruly because of neglect and had to be uprooted. As the severity of regional pandemic conditions began to wane, a couple of incarcerated gardeners in this facility were able to cultivate a new garden, settling young native plants into the soil according to a layout that was designed remotely via correspondence among various program participants.

In summer 2021, after a lapse of fifteen months, California state prisons resumed in-prison programming. Insight Garden Program staff and participants began the gradual, emotional process of reuniting with one another and restoring their gardens. While the worst of the pandemic was over, the harms caused by prison administrators' and policymakers' inability to provide care for the people in custody remained persistent and profound. And yet, the otherwise of Benjamin's "what is *possible*" continued to form the foundation for organizations and communities working toward decarceration and abolition. Indeed, for more than a year prior, and despite the trauma and fear produced by prolonged lockdown and the specter of COVID-19, the gardens remained a source of inspiration for enlivening mobilization and soul-soothing imagination. Like the community described by Gumbs, members of the Insight Garden Program community continued to center the life-generating work of gardening to recall and organize around shared needs and values. They shared a commitment to building accountability and trust despite the inevitability of conflict and the alienating effects of their institutional surroundings. And they committed to this work even though there was no way of knowing what would become of their gardens or what they would manage to grow.

Notes

1. These two laws are "the Street Terrorism Enforcement and Prevention (STEP) Act (1988) and Proposition 184, the 'three strikes and you're out' law (1994)." Gilmore, *Golden Gulag*, 6.
2. Wang, *Carceral Capitalism*; Benjamin, *Captivating Technology*; Story, *Prison Land*; Madley, "California's First Mass Incarceration System."
3. Gilmore, *Golden Gulag*; Moran, Turner, and Schliehe, "Conceptualizing the Carceral in Carceral Geography."
4. Gilmore, "Abolition Geography"; Gilmore, *Abolition Geography*.
5. Gilmore, "Literature for Justice," 1:21:48.
6. A return to abolitionist discourse and organizing emerged in response to the rapid increase in policing and imprisonment of poor people and people of color during mid-twentieth-century resistance movements and urban rebellions. Ashley Hunt describes the reemergence of abolitionism as also linked to the founding of Critical Resistance and the numerous conferences they have organized since 1998. Hunt, "Art, Abolition, and the University."
7. Kelley, "What Did Cedric Robinson Mean by Racial Capitalism?"
8. Celermajer et al., "Justice through a Multispecies Lens."
9. Taylor, "Emerging Movement."
10. Davis, *Are Prisons Obsolete?*; Costanza-Chock, *Design Justice*.
11. For instance, Kim TallBear ("Why Interspecies Thinking Needs Indigenous Standpoints") notes the absence of Indigenous knowledges represented in inter- and multispecies approaches. Janae Davis et al. ("Anthropocene, Capitalocene, ... Plantationocene?") draw attention to questions of race and colonialism lacking in recent theorizations of plantation ecologies. Büscher ("Nonhuman Turn"), meanwhile, notes how certain narratives produced within the "nonhuman turn" are selectively ahistorical and fail to provoke effective ecological politics in late capitalism. For further reviews and critiques, see inter alia Ogden, Hall, and Tanita, "Animals, Plants, People, and Things"; Galvin, "Interspecies Relations and Agrarian Worlds"; Ives, "'More-Than-Human' and 'Less-Than-Human.'"
12. Davis et al., "Anthropocene, Capitalocene, ... Plantationocene?"
13. See Hunt, "Art, Abolition, and the University."
14. Cooper, "Patience."
15. Dave, "What It Feels Like to Be Free." Cf. Thaler, "What If."
16. Weheliye, *Habeas Viscus*.
17. Jiler, *Doing Time in the Garden*; Kaye et al., "Conservation Projects in Prison"; White and Graham, "Greening Justice"; Farrier, Baybutt, and Dooris, "Mental Health and Wellbeing Benefits"; Hazelett, "Greening the Cage."

18 On plantation and slave gardens, see Douglass, *Narrative of the Life of Frederick Douglass*, 28–29; Wynter, "Novel and History, Plot and Plantation." On prisoner rehabilitation, see Lindemuth, "Designing Therapeutic Environments"; Minton, Perez, and Miller, "Voices from Behind Prison Walls"; Tom, "Humanizing Animals." On prison labor, see Rice, "Convicts Are Returning to Farming"; Brown, "How Corporations Buy—and Sell—Food Made with Prison Labor." See also Moran, "Budgie Smuggling or Doing Bird?"
19 Thanks to Cameron Allen McKean for offering this articulation.
20 Bradshaw, "Tombstone Towns and Toxic Prisons."
21 Schept and Mazurek, "Layers of Violence."
22 Pellow, *What Is Critical Environmental Justice?*; Thompson, "Prisons, Policing, and Pollution."
23 Cf. Heynen and Ybarra, "On Abolition Ecologies."
24 Cummins, *California's Radical Prison Movement*; Jackson, *Soledad Brother*; Abu-Jamal and Davis, *Jailhouse Lawyers*; Kim, Meiners, and Petty, *Long Term*; Woodfox, *Solitary*.
25 Tense and time have been difficult to manage in the construction of this text. Prematurely historicizing this period enacts a unique form of violence. At the time of writing, the pandemic, despite localized alleviation of fear and restrictions, is far from over. Its impacts continue to resound. Given that prisons have always been sites of high risk for disease outbreak, COVID-19 will likely be a concern for incarcerated people for a long time to come.
26 Hooks and Sawyer, "Mass Incarceration, COVID-19, and Community Spread."
27 Foucault, *Discipline and Punish*; Foucault, "Of Other Spaces"; Hatch, "Two Meditations in Coronatime"; *New York Times*, "Covid in the U.S."; Douglas, "Mapping Covid-19 Outbreaks in the Food System."
28 Saloner et al., "COVID-19 Cases and Deaths"; Burkhalter et al., "Incarcerated and Infected."
29 Gilmore, *Golden Gulag*, 28.
30 Bonhomme, "Troubling (Post)Colonial Histories of Medicine." Importantly, these racial and ethnic identity categories are not actually separate.
31 Gilmore and Murakawa, "Covid 19, Decarceration, and Abolition (Full)," 32:32.
32 Boisseron, *Afro-Dog*, xiii–xx (italics in original).
33 Lunstrum et al., "More-Than-Human and Deeply Human Perspectives."
34 These reflections on abolition as multispecies justice, and the emancipatory potential of prison gardens, proceed in part through an inheritance of work by Sylvia Wynter, Katherine McKittrick, and Janae Davis and her colleagues who, among many others, write of the relationship between the plots of land gardened by enslaved people and the superstructure of the plantation. This legacy is vital to research on prison gardens, and yet it is also necessary to avoid collapsing the historical plantation into the contemporary prison. See Wynter, "Novel and History, Plot and Plantation"; McKittrick, "Plantation Futures"; Davis et al., "Anthropocene, Capitalocene, … Plantationocene?"
35 Harrison, *Gardens*, 143.
36 Watkins, "Industrialized Bodies"; Rose, *Prisoner's Herbal*.

37 Myers, "From the Anthropocene to the Planthroposcene."
38 Choy, "Distribution."
39 Mbembe and Shread, "Universal Right to Breathe," S61.
40 Mbembe and Shread, "Universal Right to Breathe," S62 (italics original).
41 Gay, "Tending Joy and Practicing Delight."
42 Haraway, *Staying with the Trouble*.
43 The psychosocial and physical benefits of gardening inform the foundation of horticultural therapy/therapeutic horticulture, which has a long history. These benefits are harnessed in a variety of contexts including elder care, mental health care, and camps for displaced people. Respectively, see Han, Park, and Ahn, "Reduced Stress and Improved Physical Functional Ability"; Cipriani et al., "Systematic Review of the Effects of Horticultural Therapy"; Millican, Perkins, and Adam-Bradford, "Gardening in Displacement."
44 Waitkus, "Impact of a Garden Program."
45 Insight Garden Program is part of a statewide coalition called the Transformative In-Prison Workgroup. Part of their work involves campaigning for partial reallocation of the nearly seventeen-billion-dollar budget of state corrections toward their mission, "to ensure that all people living in prisons have access to meaningful, high quality programs, and to accelerate the impact of recent sentencing reforms towards [the] North Star goal of decarceration." See Transformative In-Prison Workgroup, "TPW."
46 Stern, "A Garden or a Grave?"
47 Lincoln, "Notes from Underground," 132.
48 Mirzoeff, *Right to Look*; Schept, "(Un)Seeing like a Prison."
49 Puig de la Bellacasa, *Matters of Care*.
50 Davis et al., "Anthropocene, Capitalocene, ... Plantationocene?"
51 Hazelett, "Greening the Cage."
52 Brown, "Of Prisons, Gardens, and the Way Out," 84.
53 By the end of March, the California Department of Corrections and Rehabilitation announced, "Inmates allowed alcohol-based hand sanitizer in approved areas under supervision."
54 California Department of Corrections and Rehabilitation, "Updates."
55 Lunstrum et al., "More-Than-Human and Deeply Human Perspectives."
56 California Department of Corrections and Rehabilitation, "Population COVID-19 Tracking"; California Department of Corrections and Rehabilitation, "CDCR/CCHCS COVID-19 Employee Status."
57 Egelko, "San Quentin Coronavirus Outbreak."
58 Sulek and Woolfolk, "Coronavirus."
59 California State Senate, "Senate Public Safety Committee." Since this initial outbreak, larger outbreaks in the prison system occurred at facilities in Avenal, Corcoran, Soledad, and San Luis Obispo.
60 Sulek and Woolfolk, "Coronavirus."
61 Hatch, *Silent Cells*; Coverdale, "Caring and the Prison."
62 Adnan Khan (@akhan1437), "There's a Chaplain who works at an outside hospital where they have been relentlessly bringing in #COVID19 patients from San

Quentin," Twitter, July 2, 2020, 3:04 p.m., https://twitter.com/akhan1437/status/1278766494176825344/. Withholding an incarcerated person's health status from next of kin is a standard practice at state and federal prisons nationwide.

63 Schlanger, "Plata v. Brown and Realignment," 166.
64 Prison Law Office, "PLO's Efforts to Address COVID-19 in California Prisons."
65 Reports indicate that, as of December 31, 2020, the total in-custody population for the California Department of Corrections and Rehabilitation was 95,432. This constitutes a more than 23 percent decrease in one year. It remains unclear as to how many of these people were released, rather than transferred elsewhere, such as to county jails or federal Immigration and Customs Enforcement facilities. The prison population is projected to increase as pandemic restrictions lessen. See California Department of Corrections and Rehabilitation, "Monthly Total Population Report Archive."
66 Legislative Analyst's Office, "The 2021–22 Budget"; California Department of Corrections and Rehabilitation, "Prison Closure Information"; Howard et al., "The People's Plan for Prison Closure."
67 Sudbury, *Global Lockdown*; Gilmore, *Golden Gulag*; Richie, *Arrested Justice*; Murakawa, *First Civil Right*; McDonald, foreword to *Captive Genders*; Schept, *Progressive Punishment*; Davis, *Freedom Is a Constant Struggle*; Wang, *Carceral Capitalism*; Platt, *Beyond These Walls*; Thuma, *All Our Trials*.
68 Movement elder and guiding light Ruth Wilson Gilmore has repeatedly offered the reminder "Abolition is about presence, not absence."
69 Taylor, "Emerging Movement."
70 Gumbs, "Freedom Seeds," 145.
71 Gilmore and Murakawa, "Covid 19, Decarceration, and Abolition (Full)," 1:10:10.
72 Benjamin, "Racial Fictions, Biological Facts," 2 (italics original).
73 Mercado, "Not Just a Gardening Program."

5

Justice at the Ends of Worlds

Michael Marder

5.1 (*previous page*) Original drawing by Feifei Zhou.

ONCE UPON A TIME, philosophy expressed its drive toward total annihilation with utmost clarity and breathtaking piety. Although the formulation in question cropped up and continued developing in the German-speaking areas of Central Europe, it was written in the language of a bygone world, namely, in Latin. And it opened a door to the perishing of the world—not only of a world belonging to a certain language or culture, a finite texture of significations, historical periods, or species, but of the world as such and as a whole.

Fiat iustitia, pereat mundus.

We might take solace in the assertion that the world as such and as a whole does not really exist and has never existed, except in the metaphysical imagination. Still, a crucial trait of this imagination has also been to give itself a perverse body in reality, to realize the dreams of metaphysics in a living nightmare. The one world of an empire, later on reworked into that of globalization, makes the prospects of a thoroughgoing world destruction eerily plausible. The underside of its unity is the ease of undercutting the world all at once, in one fell swoop. The most surprising feature, perhaps, is that this deadly activation of an incredibly effective imagination can proceed in the name of justice.

I have already placed the Latin expression, which is my subject here, in the text, suspended between paragraphs and shorn of references to its source. I have done so because there are too many sources to cite and none at all to be pinpointed as the original. The genealogy of the phrase *Fiat iustitia, pereat mundus* ("Let justice be done, even if the world perishes") is as murky as it is fascinating. As Hannah Arendt notes in one of her essays gathered in *Be-*

tween Past and Future, Ferdinand I was the likely author of the adage, and she has this to say: "Apart from its probable author in the sixteenth century (Ferdinand I, successor to Charles V), no one has used it except as a rhetorical question: Should justice be done if the world's survival is at stake? And the only great thinker who dared to go against the grain of the question was Immanuel Kant, who boldly explained that the 'proverbial saying... means in simple language: "Justice shall prevail, even though all the rascals in the world should perish as a result."'"[1] Arendt testifies, over and above her stated intentions, that the history of the dictum is a story of its distortions, whether in its all-too-free translation from Latin by Kant or in Arendt's own certainty that, besides Ferdinand I, no one risked using the sentence in the affirmative. Let's try to follow the vanishing zigzag of the saying, which promises (or threatens with) the advent of justice at the end of the world.

As in a court of law, where material evidence has to be presented before the accused is convicted or acquitted, I would like to produce the first document, a letter, written by Martin Luther and dating from 1531, which contains the lines: "Let justice be done, even if the world perishes. For I say: Throw peace to hell, if it is to be purchased at the price of harm to the Gospel" (Fiat iustitia et pereat mundus; pacem enim ad ima tartara relegandam esse dico, quae cum evangelii iactura redimitur).[2] The second piece of evidence is also a letter, which Luther addressed in 1540 to another theologian of the Reformation, Hieronymus Weller. In the letter's closing lines, Luther states: "In short, against God we can neither do, nor allow, nor tolerate anything. Let justice be done even if the world perishes" (Contra Deum nihil possumus nec facere, nec permittere, nec tolerare. Fiat iustitia et pereat mundus).[3] The context for this harsh judgment is a denunciation of prostitution, and of brothels in particular, as "our experience under Satan." It was in a similar context that Luther repeated the phrase, later to be included in his *Table Talk*, a collection of sayings uttered at dinners at Lutherhaus (my third piece of evidence): "We must hold no relations with those who seek to set up houses of evil resort. We must resolutely repress the devil, instead of encouraging him [...] We must in no way tolerate, or even wink at, aught that is contrary to the will of God: *fiat iustitia et pereat mundus*" (723).[4]

So, besides Ferdinand I who elevated the expression to his Wahlspruch, or motto, at least one of his contemporaries dared resort to the saying in the affirmative. Remarkably, the king who would become emperor and Luther, two sworn political enemies,[5] used the same maximalist expression, which saw the entire world hang in the balance. While the question "Whose world?" hardly arose, the question "Whose justice?" was most definitely implied in their contest. Would it be the justice of God, of the Gospels as interpreted in a heterodox fashion by Luther, or justice as codified in monarchical law? Will the world come to its end in the name of God or as a result of the king's edict?

Though largely omitted, the problem of the world everywhere shadows that of justice. The classical standoff of Antigone and Creon is replayed all over again there where mundus, world, is variously understood as the created realm with its mundane, political law (the "aught contrary to the will of God") included or as the imperial realm subservient to the will of the emperor who is the embodiment of justice. Add to this Luther's belief that the success of the army led by Sultan Suleiman I, to whom the cities Buda and Pest succumbed in 1527, was "another sign that the final day of God's judgment was imminent,"[6] and you will get a full picture of the end of the world that unraveled behind the scenes of the saying.

Putting the rather sketchy history of Fiat iustitia, pereat mundus aside, it is crucial to ask what these words themselves convey. First, the world, taken as an organic whole, may perish (perire), as an animal also does. In addition to the negative finitude of its limitation in time and space by its ends or edges, the world is positively defined by its ends that lend it meaning and allow it to live. Second, and relatedly, the world receives its identity as a whole from the possibility of dying as one. Third, the world's perishing may be a consequence of doing justice, of accomplishing justice not as a distant ideal but as a force brought to bear on reality. The doing of justice, its *fiat*, would then be the limit or the point through which the world would go to its fate of perishing. Fourth, despite the theological backdrop of the Last Judgment, justice, which consents to the loss of the whole world, is inherently unjust inasmuch as it fails to discern differences among parts of the world, if not among worlds. It is, then, justice without judgment. Fifth, when the sphere here below is not safe in the face of justice, neither is the heavenly abode. A companion Latin legal phrase, usually paired with the one we have been examining, is *Fiat iustitia, ruat coelum* ("Let justice be done, even if heavens fall down").[7] When all is said and done, justice brings the earth and the sky to their demise. And it survives their devastation.

To what extent did Kant go against the grain of transforming the maxim into a rhetorical question, as Arendt has it? His translation from Latin into German in the first appendix to "Perpetual Peace" obviously modifies both the letter and the spirit of what he claims to merely translate: "let justice reign, even if the rogues in the world must perish as a whole [es herrsche Gerechtigkeit, die Schelme in der Welt mögen auch insgesamt darüber zu Grunde gehen] … The world will certainly not come to an end if there are fewer bad men."[8] Note that it is no longer the world that must pass away but a whole lot of rogues (Schelme), who dwell in it. That they must die as one is, however, a judgment, which, at the same time, spares the good and demands a death penalty for the most disparate of offences comprehended by the designation "rogue" and ranging from a little mischief and the behavior of pranksters to the acts of thieves, villains, and pure evil. Kant treats Schelm as an ideal type, where singularities taken "as a whole" (insgesamt) are generalized. (Incidentally, this is the methodological problem engrained into systems of biological classifications and, above all, the category *species*). Within the field of crude discriminations, according to which the world and those who do not qualify as Schelme are to be saved, we come across the indiscriminate approach that marked the original Latin saying.

Kant, for his part, is not quite content with his own watered-down version of justice's *fiat*. In a philosophical escalation of his interpretation, he argues that, as a sound "principle of right" (Rechtsgrundsatz), the dictum substantiates what we now call the rule of law: it "should be seen as an obligation of those in power not to deny or detract from the rights of anyone out of disfavor or sympathy for others."[9] The rogues who must perish together are the poorest of the poor and cabinet ministers, plebs and oligarchs, those on the left and on the right of the political spectrum. In other words, the principle itself is "given a priori by pure reason" (a priori durch reine Vernunft gegeben ist)"[10] and, as such, is free of the pathological, contingent factors that may skew one's judgment. But here Kant, who has been initially careful to spare the world, reveals something fundamental about the statement he mistranslates: it is pure reason, disguised as justice, that decrees the necessity of the world's perishing. Why? Could it be because the messily material reality of the world does not live up to the ideal, posited by pure reason? And could it further be that their genetic incompatibility makes reason unreasonable, as Hegel would later insinuate?

The philosophical escalation of Kant's interpretation does not draw to an end here. Following a famously antiutilitarian line of thought, he writes: "This proposition simply means that, whatever the physical consequences may be [die physische Folgen daraus mögen auch sein], the political maxims adopted must not be influenced by the prospect of any benefit or happiness which might accrue to the state if it followed them."[11] Indifference to "physical consequences" is properly transcendental; it rids judgment of concerns with the ramifications of implementing the maxim. Responsive to the precepts of pure reason, a principled adoption of the maxim abdicates its responsibility for the world, not least for the world's continued existence or cessation thereof. The very ideal of perpetual peace, which Kant tries to bolster with his interpretation, is eroded through the injunction to ignore possible "benefits or happiness" (that must include peace itself) weighed for the scenarios of following or not following the maxim. In this way, Kant's thought dovetails with that of Luther, whom we have already witnessed making this deduction from the Latin expression in one of his letters: "Throw peace to hell, if it is to be purchased at the price of harm to the Gospel."

The drive toward world annihilation participates in a more encompassing drive toward ideality at loggerheads—and altogether unmediated—with existence. From Plato to Husserl, idealist ways of thinking unperturbed by the possibility that material reality would vanish all at once have imagined and, whether implicitly or explicitly, given the seal of their philosophical approval to this possibility. With Kant, the ideality of an ideal became a normative outer limit and a promise, toward which actual existence (and human history, above all) tended, but which it could never attain. The teleological end of the ideal was, by definition, beyond the end of the world (hence, also, beyond the end of history). But, its ideality notwithstanding, justice must be done; it must be in effect if it is to retain its force, translated willy-nilly into the force of law.[12] The requisite of effectiveness is, perhaps, unique to the ideal of justice, compared, say, to beauty or truth.

In the light of a critique, which Kant failed to take to its logical conclusion, the first half of the maxim—Fiat iustitia—contains a terrible contradiction. Justice must be implemented (done) *and* it must persist in the state of an unattainable ideal. What would the implementation of an ideal look like? Rather than a back and forth of negotiations and mediations with reality, it can only entail the total devastation of all that is, of all that is incapable of living up to the ideal. The second half of the maxim—pereat mundus—is, therefore, both a precondition for and a consequence of the first: a transcendental precondition bespeaking the ideality of justice indifferent to the world and an empirical consequence of unleashing justice onto the world. It is in this sense that Hegel's *Phenomenology* invokes postrevolutionary Terror in Robespierre's France as an issue of absolute, universal freedom—first and foremost, a historically realized freedom from actuality. Since, Hegel observes, one "cannot call on any part to serve as a middle term to connect universal freedom and actuality, their relationship is that of wholly *unmediated* pure negation; namely, it is that of the negation of the individual as *existent* within the universal. The sole work and deed of universal freedom is, in fact, *death* [Das einzige Werk und Tat der allgemeinen Freiheit ist daher der *Tod*]."[13]

It is worth mentioning, in a restatement of a point I've made earlier, that the devastation of the world by an ideal, be it of freedom or of justice, becomes or gives itself a world. Something survives the end of the world; something outlives the "sole work and deed" of an abstract universal, the work and deed that is death. World destitution *is* the world seen from the perspective of metaphysics, just as, in the dialectical scheme of *Phenomenology*, the stage of "absolute freedom and Terror" is not the end as the decisive cessation of the world; it is followed, across the transformative gap of mortality, by "the moral worldview." Indeed, the section on morality in Hegel's *Philosophy of Right* is where the German philosopher openly criticizes whoever endorses the Latin maxim. Having defined the good in terms of "*realized freedom, the absolute end purpose of the world* [die realisierte Freiheit, der absolute Endzweck der Welt]," he writes in the following paragraph: "Wellbeing is not good without right. Similarly, right is not the good without wellbeing (fiat iustitia should not have pereat mundus as its consequence [Folge]). Thus, since the good must necessarily be actualized [wirklich zu sein] through the particular will, and since it is at the same time the latter's substance, it has an *absolute right*."[14]

If the world has an end, that end is, in keeping with Hegel's thinking, the realization of freedom and, therefore, a new beginning—the absolute beginning, one might say. Unlike abstract universal freedom, the realized freedom he invokes does not demand the sacrifice of existence as such and as a whole to the Moloch of an ideal. The end is nothing more and nothing less than freedom actualized in the world *as that world*. In parallel, the domain of right (that is, of justice) is vacuous without the actualized good, which, rather than a promise, has been fleshed out in specific activities and their outcomes. For this reason, acts of justice cannot be followed by the perishing of the world: the doing (fiat) of justice, its becoming-actual, is the becoming of the world itself at its "absolute end purpose." Only a pure negation, engrained into universal freedom, is capable of envisaging the devastation of the world as its ultimate result.

How should we grasp the becoming-actual of justice, which seems to be in tandem with the becoming-actual of the world? According to the reading protocols I have established in my *Energy Dreams* and, subsequently, in *Hegel's Energy*,[15] *actuality* (Wirklichkeit) is Hegel's word for energy in a complex dialectical attempt to inherit and transform Aristotle's coinage, energeia. Whereas, in the modern mindset, energy (much like the ideal of justice) essentially stands for a promise, a potentiality that can never be actualized, for Aristotle, it is precisely the other of potentiality, or, in his words, not-dunamis: energeia "means the being of the thing, not in the sense which we mean by 'potentially' [ἔστι δὴ ἐνέργεια τὸ ὑπάρχειν τὸ πρᾶγμα μὴ οὕτως ὥσπερ λέγομεν δυνάμει]" (*Met.* 1048a, 31–32). The becoming-actual of justice and of the world is, far from entropy or senescence setting in, an experience of energetic fullness, of a promise fulfilled.

Lest you be tempted to read utopian overtones into these formulations, consider the implications of interpreting energy in the terms and on the terrain of actuality. It has nothing to do with attaining an ideal in a predetermined teleological journey, in the course of which bare potentialities are actualized. Energy is the being of a thing (of the world, of justice) at work and in the work (ergon). Now, justice can be neither at work nor in the work outside the world; in that hypothetical, extramundane sphere, it can exist only as a pure possibility, an ideal, or, in Aristotle's words, as dunamis, not energeia. The energy of justice is also that of the world. The Latin dictum ought to be rewritten as Fiat iustitia, floreat mundus, "Let justice be done, so that the world flourishes."

The world's preservation through the work of justice is not the sole divergence from the genealogy of the maxim we have been tracing on these pages. You might recall that in the original phrase, the world perishes as a whole, an immense animal superorganism that coincides with the Earth (Gaia) as a living system, dying in an instant. From its mortal edge, ending as a result of justice's fiat, the world receives its identity and final determinacy. But, assuming that an act of justice can only promote the most varied forms of the world's wellbeing, this unity of an end and, hence, of the world itself is no longer a given. The flourishing of the world bursts out into world*s* that, though still finite, have their ends defined by the energies-actualities at work and in the work in them. They are vegetal, not animal—even when animals are the ones flourishing. It follows that the world does not die all at once, for it does not live all at once (at most, different lifeworlds partially overlap in the same objective space and time). The energies of justice mark the edges of each world, from the species all the way down to nongeneralizable singularities, their burgeoning, flourishing, or flowering happening when they come into contact with one another. From these energies, worlds receive their nonidentity.

We should linger for some time with this notion of worlds so as to appreciate its interaction with the no longer ideal work of justice. What I have in mind is an amalgam of Jakob von Uexküll's idea of the Umwelt (literally: the enveloping or surrounding world; more loosely: an environment);[16] Heidegger's being-in-the-world as the existential condition of meaning making or sense generation, as well as the divergence between Earth and world in his late philosophy;[17] and Derrida's singularizing identification of each biography with a world.[18] According to this combined theory, which also has a lot in common with Nietzsche's perspectivalism and Eduardo Viveiros de Castro's multinaturalism, worlds are the worlds of sense, making sense from within to those whose worlds they happen to be. It would be as ridiculous at best and harmful at worst to invoke one world as to argue that one structure of sense or a single pattern of meaning is equally applicable to all existences. Justice in the workings of distinct, if partially overlapping and mutually touching, worlds entails their adjustment to one another, the attunement of one sense or set of meanings to another. What would be ruinous, then, would not be justice but its opposite, namely, injustice as heedlessness to a different sense, with which a being comes into contact.

The multiplicity of worlds supplanting *the* world challenges the colonial legacy, not least on the most literal of registers. The colonial world (colonia) was land to be settled and cultivated, shaped according to the desires and requirements of the settler (colonus). Indeed, from the colonial perspective, it did not have the character of a world prior to the acts of settling and cultivation by those who deemed themselves the first to have reached its shores or trodden its ground. In the eyes of the colonizers, the colonial world prior to their arrival was not a world at all; it merged with the earth, its edge standing for the temporarily unknown, a mobile frontier. According to the precept of ideal justice, such a world might as well perish, its end freeing up the innumerable world of colonized human and other-than-human beings. The difficulty, however, is that what would perish in this case would not be a world but, on the contrary, a structure of containment that holds worlds at bay. The flourishing of decolonial worlds would then be empowered by the perishing of the colonial not-quite-world.[19]

The site of another divergence from the lethal ideal of justice is the subtle distinction between force and energy. As an ideal, justice relies on brute force to be foisted onto the reality it, at the extreme, destroys. This force is external, and it is deployed to enforce the impossible compliance of the real with the ideal. The energies of justice, for their part, are without force, even though they are the precondition for its mobilization. They are the actualities of various worlds taken *just as they are*: plural, communicative, essentially imperfect, finite. Consequently, the doing of justice that hones its energies is a practical edge, at which a world goes to its fate of encountering other worlds.

Understood in this way, justice is much more common than we tend to think; rather than an unreachable ideal, it is what happens when there is coexistence between disparate worlds that share, for a time being or on a more permanent basis, the same physical space. As what *just is*, justice may be inferred from the mutual adjustment of worlds, intercalated at their edges. Whereas, for Darwin, evolutionary fitness had to do with the physical and physiological adjustment of a species to its home environment, the fit of justice that transpires between worlds is no longer between a kind of being and its more or less backgrounded milieu. It happens, instead, between two or more dynamic structures of sense that are able to maintain themselves as worlds precisely thanks to the interactivity and interpassivity of their interworld. Obviously, what *just is* does not refer to a single entity but to pluralities that, in order to keep existing and flourishing in all their diversity, have no other choice but to practice mutual fittingness and continual adjustment to each other. There is, also, no single measure of having succeeded in this endeavor that is coextensive with life itself (or, better, with lives). Nonetheless, extinction (and, particularly, mass extinction) is an ersatz barometer of injustice and failure, where what *just has been* no longer is.

Like Fiat iustitia, pereat mundus, the energies of justice refrain from passing a formal judgment on the world or on worlds. That said, the results of this refraining are diametrically opposed in the two cases: the salient features of the first are metaphysical indifference toward and lack of differentiation in the world condemned to perish as a whole; in the second, there is minute, precognitive, ontological discernment that renders to each world its own position, mode of opening unto other worlds, and time. To reiterate, the flourishing of worlds, which the energies of justice both foster and express, is not a final, postapocalyptic victory over perishing and, therefore, death. The worlds of various species (not to mention of singular beings) are finite, and the contacts taking place between and among them, at their edges, tip the scales one way or another: toward exhaustion and extinction, or toward the mutual support of fragile existences. The energies of justice surge to the extent that this support is practiced, but they start growing already when another world is accepted just as it is with the attitude of a profound non-indifference. They germinate in caring *about* worlds and come to fruition in caring *for* worlds.[20]

In the prismatic light of the energies of justice, what emerges is an ethical phenomenology of worlds, not an ideology inimical to the world's continued existence. Everything that is, just as it is, can be neither comprehended nor described with a single theoretical gesture; and, on the contrary, what is not and can never be—namely, justice as an ideal set over and against the world—transforms that world after its own image into what will not be, what will perish as a whole. The fact that Hegel sides with the energies of justice against this murderous *fiat* indicates that his is not a totalizing philosophy but one that embraces and justifies the right to be of all that is.

What sort of multispecies justice is possible in the energetic embrace of what just is? I much prefer the term *multiworld* over *multispecies*, which is still redolent of the hierarchies inherent to taxonomies and systems of classification. The lines of demarcation among different species may certainly become permeable, shifting with time, or they may be inherently open to contestations even when it comes to the criteria of membership, from morphological to phylogenetic.[21] But, however "promiscuous" the realism that stems from the acknowledgement of "vague boundaries" between species,[22] it remains a *realism*, that is to say, a set of assertions about things, about the real, that does not proceed from the standpoint of these things themselves. At bottom, the dispute between *species* and *world* is a contest between realism and phenomenology, which invariably commences from and returns to the perspective of the matters themselves under consideration. The question is one of malleable boundaries experienced *from within*, more as edges of worlds than separations between species. Perhaps the closest that species-thinking can come to phenomenology and worlding is through praxiography, according to which "species emerge amidst intra-actions with companions."[23] Such "intra-actions" are the interworld engagements, by virtue of which something like species is possible in the first place.

Nor does the multiplicity of multiworld or multispecies imply a strictly quantitative category. There is no standpoint outside an array of worlds, from which to overview all or even some of them; it is only the ideal of justice that feigns such a standpoint in preparation for its fiat. Though ostensibly opposed to world destruction and committed to what just is, the value of preserving biodiversity is the other side of this counterfeit metaphysical coin. Lifeworlds are incalculably more than useful gene pools, those resource and information banks on which to draw in the future. In effect, translated into the barebones of DNA, the actual carriers of genetic information become dispensable: it makes no difference whatsoever if they perish or not, so long as the information itself is saved. Concentrating on what just is requires us to rid ourselves of an obsession with the deep structures and inner potentialities of being, focusing instead on the surfaces, materialities, and actualities of existence that are crucial to a phenomenological outlook.

While not situated outside the world, multispecies or multiworld justice is not intramundane, either. It concentrates on the spatiotemporal edges, at the ends of worlds that, far from serving as the markers of determinacy apt for an epistemological demarcation of species boundaries, bristle with the promise of the event of an encounter with other worlds. A consequence of Fiat iustitia, flourishing (florire) takes the place of the world's perishing (perire). In a paraphrase of Freud, where the perishing was, there the flourishing shall be. So, what changes between what was and what shall be? To perish is to go through and beyond the end to nothing—to the nothing that surrounds the one world on every side. To flourish is also to go through and beyond the end of a world to another world—to other worlds that border and touch on it. Flourishing happens among worlds, between their dispersed ends and beginnings, in the unending middle of finite existence. We must, henceforth, learn perishing, or going through to our ends, otherwise than in the way that nihilistic metaphysics has taught us: to twist perishing into flourishing. And this world pedagogy is the work of justice.

Notes

1. Arendt, *Between Past and Future*, 224.
2. Luther, *Briefwechsel*, 76.
3. Luther, *Letters of Spiritual Counsel*, 292.
4. Luther, *Table Talk of Martin Luther*, 302.
5. In 1523, just two years after taking the rule over traditional Hapsburg areas, Ferdinand "declared his intentions against Luther and his ally Zwingli and in the next year revealed that he wanted to go after them." Marty, *Martin Luther*, 165. That said, in the 1555 Declaratio Ferdinandei, the principles of a peaceful coexistence of Catholics and Protestants in the Holy Roman Empire were set into law.
6. Marty, *Martin Luther*, 164.
7. Most famously, this expression was used in British common law in the 1772 Somerset case, in which "James Somerset claimed that he was a free man, and that Charles Stewart could not enforce his title to Somerset as a slave owner, despite having purchased Somerset while in Boston." McGaughey, *Casebook on Labour Law*, 9. Nonetheless, the origins of this expression are as obscure as those of its more commonly used counterpart. According to Thomas Bayly's Royal Charter, at a trial in Geneva, it is "Calvin alone who stands up and cries fiat iustitia, ruat Coelum." Bayly, *Royal Charter*, 131. Be this as it may, the positions of Creon and Antigone may be now allocated to these two distinct expressions: ruat coelum to the former; pereat mundus to the latter.
8. Kant, "Perpetual Peace," 123–24.
9. Kant, "Perpetual Peace," 123.
10. Kant, "Perpetual Peace," 124.

11 Kant, "Perpetual Peace," 123–24.
12 I cannot resist the urge to mention that one of the variations on the fiat of justice is Fiat ius, ruat iustitia ("Let law be applied, even if justice collapses"). Consult Rapalje and Lawrence, *Dictionary of American and English Law*, 2:513.
13 Hegel, *Phenomenology of Spirit*, 343 (emphasis original). Translation modified.
14 Hegel, *Philosophy of Right*, 157–58. Translation modified.
15 Marder, *Energy Dreams*; Marder, *Hegel's Energy*.
16 Uexküll, *Foray into the Worlds*.
17 Heidegger, *Being and Time*.
18 For a relevant thesis, see the original French title of *The Work of Mourning*, namely, *Chaque fois unique, la fin du monde* (Each time unique, the end of the world).
19 In effect, the original members of the postcolonial (or subaltern) studies circle have contributed a great deal toward the task of deconstructing the unified and monumental notion of "world." Having consolidated this work in her *Critique of Postcolonial Reason*, Spivak extends her insights further in *Death of a Discipline*, where she opposes hegemonic globality with a much more welcoming and fecund planetarity, accommodating a plurality of worlds. "If we imagine ourselves as planetary subjects rather than global agents," she writes, "planetary creatures rather than global entities, alterity remains underived from us; it is not our dialectical negation, it contains us as much as it flings us away" (Spivak, *Death of a Discipline*, 73). In *Provincializing Europe*, another member of the original subaltern studies circle, Dipesh Chakrabarty, shifts the emphasis from the world as a projection of Europe to what he calls "non-European life-worlds": "My purpose is to explore the capacities and limitations of certain European social and political categories in conceptualizing political modernity in the context of non-European life-worlds. In demonstrating this, I turn to historical details of particular life-worlds I have known with some degree of intimacy" (20). More recently, Chakrabarty has turned his attention to the question of climate and, in his words, "planetary histories." In *The Crisis of Civilization: Exploring Global and Planetary Histories*, Chakrabarty shatters the world into worlds, showing that globality is not only a Eurocentric but also an anthropocentric construction, which ought to be replaced with *planetarity* (a term he shares with Spivak) that opens onto other-than-human worlds. In a parallel vein within decolonial studies, Walter Mignolo and Catherine Walsh (*On Decoloniality*) seek the promise of worlds in the "fissures or cracks [that] are present throughout the world, including in the Global North." Decolonial worlds flourish out of these fissures or cracks, including the "new movements, philosophies, and horizons of and for praxis" that open up in the South—"the South in the South and the South in the North" (24). And, just as postcolonial theorists have moved beyond an anthropocentric perspective, so have scholars working in the field of decolonial studies. The latest examples of this tendency include Gómez-Barris, *Extractive Zone*; Cabot, *Ecologies of Participation*; and a recently published book, Cooke and Denney, *Transcultural Ecocriticism*. Similarly, multidisciplinary researchers have creatively brought together post- and decolo-

nial studies, on the one hand, and nonanthropocentric perspectives, on the other. In this regard, consult Parreñas, *Decolonizing Extinction*.
20 For this distinction within care, see Marder, "We Couldn't Care More!"
21 Dupré, *Disorder of Things*, 44.
22 Dupré, *Disorder of Things*, 36.
23 Kirksey, "Species," 776.

6

from the micronesian kingfisher

Craig Santos Perez

6.1 (*previous page*) Original drawing by Feifei Zhou.

~

hasso' in elementary school
[we] color pictures of
the micronesian kingfisher [sihek]

black beak
blue tail
green wings
orange & white feathers

a native bird [we]
never saw in the jungle

~

american zookeepers
arrived in the 1980s
wrote a species survival plan
captured 29 of the last wild
[sihek] on guam
& shipped them
    off-island to safety

*1944: the first brown tree snakes
arrive on guam as stowaways
aboard us military ships carrying
equipment & salvaged
war material from military bases
in papua new guinea*

~

[we] memorize & recite
its scientific name
    halcyon cinnamomina
cinnamomina

~

enclosure :
exterior features :
quarter inch plywood
screened mesh front
interior : ceiling foam rubber
or burlap stuffed with straw
external minimum size :
nine inch by nine inch
internal height :
ten inches
between floor
& ceiling padding

*1953 : first written evidence
of brown tree snakes
in the apra harbor area
the most likely introduction site
on guam
the sightings of snakes
are referred to as rumor*

cage size
for breeding pairs :
ten feet by eight feet by ten feet

from the micronesian kingfisher

~

[we] practice spelling
"endangered" & "extinct"

1960s: snakes
are reported throughout
southern & central guam
by civilians & members
of the military

learn prefixes
"ex-" & "en-"

study for
the vocabulary quiz

~

eve-
ry
ch
-ick
is
ex-
treme
-ly
pre-
c
(ar)
-ious

what follows your flag

~

[we] construct a fake nest in class

with twigs from the playground
torn paper & glue

[we] place plastic
easter eggs in the center

~

[2010]
"a mated pair of micronesian kingfishers laid two fertile eggs this spring
inside a hollowed-out palm log in a special breeding room of the chicago
lincoln park zoo bird house   the parents incubated & hatched one egg
in the hollow log   keepers took the other egg which hatched a few days
later inside an incubation machine in a lab which mimics the conditions
of a nest"

~

"the micronesian kingfishers are
            notoriously difficult to breed
        & they don't always properly
            care for their eggs"

from the micronesian kingfisher

~

        during incubation
        keepers
        track the chick's
        development
        by shining a light
        through
        the eggshell

1968: snakes
have spread
throughout the island
are confirmed
at the northernmost area
of ritidian point

        an act known as
        "candling"

~

after 20 days
the tiny egg
begins wiggling
& cracking
until the chick
emerges
at the smithsonian
conservation
biology institute
in virginia
[2019]

1981: a brown tree snake is found in the customs area of the honolulu airport having hitched its way from guam. a second snake is discovered near an aircraft hangar at barbers point naval air station

      less than
an inch long
8.5 grams
in weight
pink
featherless
female

1970s-80s: the snake population continues to grow exponentially with reports coming in from across the island declining bird populations are noticed

from the micronesian kingfisher

~

a camera
inside
the incubator
captures
the moment
the chick
hatches

you can view
the video here:

https://nationalzoo.si.edu/news/extinct-wild-guam-kingfisher
-hatches-smithsonian-conservation-biology-institute

~

the chick
is born
with eyes
closed

~

without birds
thousands of native
black butterflies
emerge from cocoons
all at once *[2019]*

*1984: most native forest birds
on guam were virtually extinct
when they were listed
as threatened or endangered
by the us fish and wildlife service*

they search for nectar
in flower blossoms on trees
at the war in the pacific
national historical park

from the micronesian kingfisher

~

[we] draw colorful timelines on poster board

day 1 : blind & naked

day 2 : casts produced

day 5 : flight feather tracts visible on wings

day 7 : feather tracts visible on back sides & head

day 10 : eyes begin to open

day 13 : feathers begin breaking through skin

day 19 : breast feathers breaking from sheaths

day 20 : skin completely covered by pin feathers

day 27 : feathers emerge from sheaths

day 29 : perching

day 30: fully feathered

day 35 : fledging

~

"do the birds eat spam"
i ask our teacher

she laughs & says
the mom & dad [sihek]
feed the chicks
insects lizards
crabs & shrimp

1993: a brown tree snake
makes its way
to continental us
for the first time
discovered in a crate
of household goods en route
from guam
delivered to inegside naval station
on the north side of
corpus christi bay in texas

~

keepers feed
the chick
every two hours
between 6 am & 6 pm
thawed mice
mealworms
crickets & lizards

keepers feed
the chick
with tweezers
protruding beneath
the beak of
an oversized kingfisher
hand puppet

from the micronesian kingfisher

~

"guam has forty times
more spiders
than other islands"

*1986: a brown tree snake
is discovered on a ship
carrying naval cargo
as it came to anchor
off the island of diego garcia
because this is a major
military base in the indian ocean
there is strong probability
this snake was a stowaway
from a stopover in guam*

spiderwebs
weave
every foot
of the jungle

"there's no other place
in the world
that has lost all
its insect-eating birds"

~

keepers
place mirrors
in the incubator
& play [sihek]
vocalizations
to prevent
the chick
from imprinting
on humans

~

[we] listen
to an audio tape
of [sihek]

[we] mimic
its song

"kshh-skshh-skshh-kroo-ee kroo-ee"

from the micronesian kingfisher

nativ

       tre s

             disap-

                thin
              -ni   n

        ope   n

can

       -opy

in 2010
the us department of agriculture
"bombed" the island   fract-
with dead frozen mice
laced with acetaminophen   g ps
which are lethal to snakes

  un-
   ger   m
                -inat d

                     see

                        -ds

~

"guam is
a natural laboratory"

"there's no other place
where you can see
what happens when birds
are removed from
an entire ecosystem"

~

"approximately 140
micronesian kingfishers
are alive today
all descended from
the last 29 wild birds
taken into captivity
in the 1980s"

"all in human care"

from the micronesian kingfisher

~

*will guam
ever be safe
for [sihek]
to return
home*

~

a male [sihek]
died today *[2017]*
at the smithsonian
national zoo's
bird house
in washington dc

he was 17 years old

*the us department of agriculture
traps 6,000 brown tree snakes each
year yet there are still nearly
two million on the island
the most dense patches contain
14,000 snakes per square mile
one of the highest snake
concentrations in the world*

~

*will guåhan
ever be safe
to re-wild
native
birdsong*

~

i see a living [sihek]
for the first time
at the san diego zoo

i am 20 years old *[2000]*
attending college in california

"hafa adai"
i whisper

into the cage

"guahu si craig
familian gollo

ginen mongmong
but I live here now

like you"

from the micronesian kingfisher

avian silence

"let me tell you
about guåhan ..."

7

Rodent Trapping and the Just Possible

Jia Hui Lee

7.1 (*previous page*) Original drawing by Feifei Zhou.

IT IS A COMMON SCENE IN TANZANIA: On bicycles, at the market, and all around the bus station, men sell rat poison, glue boards, live traps, and snap traps. Some of these are imported from China, and others from Oman. Whether in the business capital of Dar es Salaam or in smaller cities like Morogoro, people talk and complain about *panya*, a Kiswahili term meaning "rodent" and encompassing all gnawing small mammals. Panya figure in stories about children getting bitten, missing articles of clothing, denuded corncobs, and occasionally witchcraft, all told by Tanzanians who purchase rodent control technologies in the hopes of keeping rodents out of their homes and fields.

Iddy Juma Kilongola is a man in his mid-forties who designs and makes rodent traps, which he then sells at the weekly Saturday market (see figure 7.2) in Morogoro not far from the Sokoine University Pest Management Center.[1] His stall features several small kill traps, which resemble a box fitted with a spring mechanism. There are also larger traps made of thin steel, woven in the shape of a lantern. These contraptions are used for live trapping. Pointing to a coffin-sized trap filled with scurrying, squeaking mice, Iddy boasted with a salesman's shrewdness, "This can catch four hundred in one day." My curiosity was piqued by the array of traps he had fashioned—and particularly by those designed to catch rodents alive. "Why would you want to catch them alive?" I asked. "Because," Iddy chuckled, "in the countryside, panya are a snack [mboga]."[2]

In our interviews, Iddy often spoke about his traps as technological inventions that offered more just modes of rodent control over other commercially produced traps and rodenticides. In this chapter, I explore how Iddy and

7.2 Iddy at his rodent trap stall and workshop in Morogoro town center. Photograph by Jia Hui Lee.

other inhabitants of Morogoro struggle to earn a living that often requires meditating on practices of killing rodents. Central to these practices of attaining "the good life" is the imagination of the *just possible* that manifests materially in the design and use of rodent traps. In an agricultural region of Tanzania where rodent abundance constantly threatens human sustenance, the just possible comes to life through Iddy's situated trap making, capturing the many ways in which Tanzanian selves and communities successfully produce sustenance in a world where it is seemingly scarce. Visions of the just possible like Iddy's trap-making endeavor, as we shall see, generate new social relations and forms of value that in turn hold unexpected promises of multispecies justice.

Commensal rodents—that is, rats and mice that share food with humans and are thus usually copresent with people—are widely regarded as pests. The figure of the pest occupies an ambiguous position within discussions of animal welfare and ecological conservation.[3] Most ecologists and proponents of animal welfare agree that the prioritization of certain ecosystems and economies often justifies control of animal populations that threaten conservation goals or business.[4] Debates concerning humaneness as it pertains to various pest management methods are in fact conversations about killing. *Humane*, as an adjective used by practitioners of animal population management, signals how swiftly, efficiently, and painlessly animal pests are

Rodent Trapping and the Just Possible

killed or removed through culling, trapping, or poisoning.⁵ The designation of the term *pest* itself implicates political valuations that strike at the heart of what Achille Mbembe calls "necropolitics," or "the power and the capacity to dictate who may live and who must die."⁶ When considered within the lexicon of pest management, necropolitics makes visible who has power to influence policy decisions, whose livelihoods are worth protecting from the threat of pests, and—in the context of colonialism and racism in Africa— which human and nonhuman lives are deemed extinguishable. When considered within historical and social contexts of human-animal relations in Africa, the term *humane* as it is understood in terms of pest management does not just center on questions of how humans should treat other animals but also who is afforded the dignity of being human. This double meaning of *human/e* threads through Iddy's traps, as well.

In the following, I show how we can interpret Iddy's traps as instruments both of killing and of justice. The very practices of making technologies for rodent killing are, through Iddy's creative designs, meditations on the limits of animal welfare as a framework for discussing pest control. Discussions about animal welfare can sometimes overlook those precarious livelihoods that depend on minimizing the effects of "animals that cause damage" (wa-haribifu), the Tanzanian expression for "pests." Rather than focus on the term *human/e* and its accompanying ethical considerations of what killing should look like, I hope to underscore how Iddy and other Tanzanians articulate a version of multispecies justice through their efforts to preserve their livelihoods amid challenges posed by rodents, a lack of resources, and limited formal education. Through creative design, Iddy's artisanal traps are deeply informed by a desire to improve the lives of his community while struggling with the ethics of killing rodents. I present Iddy's traps as crucial material-semiotic interventions into discussions about multispecies relations in contexts where human lives exist at the very edge of survival.

Beyond multispecies considerations, I also hope to interrogate notions of justice in relation to decolonizing scholarship on African technology. The Kiswahili term *fundi*, or fabricator, succinctly captures Iddy's ability to assemble and mobilize skills, experimentation, and social relations to bring his traps to life. In attending to the intellectual and physical labor entailed by fundi like Iddy, I seek to recuperate African technological endeavors that are often written out of global histories of technology. In doing so, I follow Kenda Mutongi's call for scholars to "take seriously what ordinary Africans are making in Africa and how they are making it."⁷ To this end, I approach Iddy's artisanal repurposing of construction materials and techniques as a form of inventive and intellectual labor. Through his technical mastery of trap making, Iddy imagines and generates just-possible futures, whose subjects include not just Iddy and his community but also the intended target of Iddy's traps—rodents.

The "First Robot": Traps as Intellection

At the workshop located opposite the daladala (minibus) stand, Iddy displays his traps under an umbrella on a reused Vodacom advertisement banner amid piles of wood and metal spokes. He is usually seated on a machine he calls a goat (mbuzi), which he invented to drill holes through wood pieces and conjoin them into traps. Iddy painted the so-called goat and decorated it with the slogan "Tanzania ya Viwanda," meaning "Industrial Tanzania," which invokes the government's development plan to build up Tanzania's manufacturing economy. Under this he added "Ubunifu Kwanza" (Imagination/Invention Comes First). In Kiswahili, *ubunifu* simultaneously refers to imagination, creativity, and invention, all qualities that Iddy's enterprise embodies. In our conversations, Iddy expressed hopes for someday operating a "trap factory" that would provide economic opportunities to farmers and youth, many of whom struggle to find gainful employment in Tanzania. Placing himself squarely within the nation's industrializing aspirations through colorful designs on his machine, Iddy dreamed that his trap factory will "bring fortune [baraka] to the whole country."

Putting together bicycle gears and leftover construction materials to design and build innovative rodent traps, one could argue that Iddy shares certain qualities with engineers and computer scientists who create and support software that is free and open source. These efforts allow anyone to distribute, modify, and make use of software without profits accumulating exclusively to owners of intellectual property. Iddy's ability to tailor his trap designs to better suit community needs, rather than commercial ones, evokes practices of designing free and open-source technologies that cybernetics scholar Ron Eglash describes as a form of engendering "generative justice." Inspired by the makers of Arduino and other open-source platforms, Eglash contrasts "generative justice" with "distributive" and "restorative justice." These latter ways, he suggests, often place demands for social justice on authorities and governments, conceding a top-down view of justice.[8]

On the contrary, generative justice emerges from the very people whose work creates value for themselves and for others in their communities through constantly shifting social arrangements. Instead of conceptions of justice that issue from questions about distribution or individualist capabilities, generative justice prioritizes social practices of living well that transform oppressive systems.[9] I consider Iddy's engineering a kind of generative justice within a multispecies community of farmers, trappers, and rodents. His traps are deeply embedded in a production process substantially shaped by interspecies relations and communal concerns, specifically suited to the needs of the more-than-human communities that Iddy inhabits.

To recognize rodent traps as instruments of justice making is not to ignore the fact that traps capture—and often kill—their prey with "unthinking, poised violence"[10] and "deliberate wickedness,"[11] as the anthropologists Alfred Gell and Lewis Henry Morgan respectively observe more than a century apart.[12] Donna Haraway notes that "there is no way to eat and not to kill."[13] In Tanzania, where agriculture subtends and supports people's ability to thrive, rodent trapping exists within a matrix of quotidian calculations for survival. To be able "to eat," which is also an idiomatic way of saying "to earn a living" in Kiswahili, depends on how much food one must share with uninvited others such as rodents. In this regard, Iddy's traps are material manifestations of how Tanzanians think ethically about killing those with whom they must share food.

Growing one's own food became an important survival strategy in the context of food rationing measures in the 1980s, when Tanzania was subject to austere structural adjustment programs. Yet even before that decade of struggle, having enough to eat had always been a key priority for many Tanzanians and the foundation for all personal development (maendeleo). "Chakula ni uhai," so the saying goes, or "Food is life." For this reason, the figure of the farmer holds a high moral position in Tanzanian society. The hard, grueling labor of farming is considered noble and associated with feeding the family and developing the nation. Agriculture has always been Tanzania's largest economic sector. Tanzania's founding father and first president, Julius Kambarage Nyerere, described agriculture as "the foundation of all our progress."[14] Thirty years later, in 2009, President Jakaya Mrisho Kikwete launched a national economic initiative, Kilimo Kwanza ("Farming First"), to modernize the agricultural sector as the nation's main driver of development. Agriculture accounted for roughly one-third of the country's gross domestic product in 2017, a figure that does not include the many food products that come from people's gardens, sometimes supplementing household income.[15]

Almost all the men and women I know in Morogoro participate in some form of agriculture. Being able to garden or farm is considered a crucial life skill. In small pockets of gardens, even close to the town center, people plant stalks of corn and cassava or tend banana and papaya groves. Often, salaries from paid work are insufficient to meet household need, so people rely on their gardens for nourishment. As Rashidi, a rodent trapper, explained when describing his 250-square-foot garden, "What we grow we don't buy. The money we save, we use to pay for our children's schooling." The room I lived in during fieldwork was part of a larger compound owned by a landlady who often shared her bounty of fruit, lemongrass, and vegetables with me. "If you don't eat them, the ngedere [vervet monkeys] will," she would say.

If food is indeed life and part of an intricate calculus for survival and success in Tanzania, then the harvesting of garden produce by nonhuman entities must be weighed up against household budgets, school fees, delayed wages, and rapidly rising costs of living. Experiences with animals that cause damage to harvests, or waharibifu, are common. These critters include grain borers and weevils, vervet monkeys, bamboo rats, field mice, and mongoose, among others. Rodents figure frequently and perniciously in local residents' accounts. They "attack" during the planting and growing seasons. They infest homes, biting children or stealing items of clothing, especially underwear (chupi). They appear without warning and in swarms. Rats and mice devour newly planted seeds and seedlings or climb up corn stalks to gobble up maturing cobs. During the months of January and February, it is common to meet a despondent acquaintance who has had an entire weekend's worth of sowing devastated overnight by a ravenous pack of rats.

Due to its mountain ranges, fertile soil, and diverse climates, Morogoro region supplies Tanzania with myriad fruits, grains, and vegetables, including strawberries, maize, rice, papaya, bananas, onions, and millet. Consequently, Morogoro town is also home to the Ministry of Agriculture's Rodent Control Centre, as well as Tanzania's only agricultural university, the Sokoine University of Agriculture (SUA). Iddy's trap-making enterprise thus stands within a society that confronts in many ways the problem of learning, in Haraway's words, how "to live responsibly within the multiplicitous necessity and labor of killing" as part of daily life.[16]

Embedded within a context where killing rodents is unavoidable, Iddy and other trappers are deeply "engaged in intellection, firmly anchored in their own philosophies, and alert to the world around and beyond them as a source of things that they render technological."[17] Striving to flourish with just enough resources in ways that foster socially just and possible futures, they practice what I call the condition of the just possible. Trap makers and farmers leave open the possibility of cultivating multispecies well-being through their experimentation with and deployment of traps. Contending with the labor of killing, they wrestle with the entangled, "emergent ecologies" that bind crops, rodents, and humans together.[18] The act of trapping rodents does not always fit within schemes for eradication and control. Sometimes, traps serve to catch food, which may include rodents. In their effort to craft just-possible futures for themselves, others, and "other others," trap makers thus complicate the simplistic and deadly designations of rodents as pests.[19]

In positioning traps as practical and theoretical tools that navigate the daily realities of living with rodents, I both invoke and challenge extant anthropological literature on traps and trapping.[20] Anthropologists have long admired the technical sophistication involved in the design of traps, often comparing their workings to electrical circuits or motherboards. Like open-source software, traps and their designs circulate freely. They are adopted, appropriated, and repurposed through dynamic processes of migration, exchange, and circulation. Edward Burnett Tylor, writing at the end of the nineteenth century, considered traps alongside other implements such as weapons and wheels as evidence of mental development among people whom he called "primitive."[21]

Other anthropologists like Julius E. Lips, who did extensive work among the Innu of the Labrador Peninsula, considered the trap to be the "First Robot," an invention that was "certainly of greater consequence to the history of mankind than the invention of the wheel."[22] Lips surveyed traps from North America, West Africa, and Europe, concluding that they are possibly "the oldest application of relay structures" and that they formed an integral part of any "modern technique" of automation and information processing.[23] In a comparable vein, Alfred Gell in his essay entitled "Vogel's Net: Traps as Artworks and Artworks as Traps" describes traps as a kind of "automaton," with a cybernetic ability to produce action in the absence of a person. Gell praises traps as devices that "embody ideas [and] convey meanings" because the trap, "by its very nature, is a transformed representation of its maker, the hunter, and the prey animal, its victim, and of their mutual relationship."[24] Posing traps as a "nexus of intentionalities between hunters and prey animals," Gell evokes their ability to bring together different worlds.[25] In other words, Gell suggests that traps are portals through which sensory worlds collide and converge. Trap designers imagine and inhabit the sensory worlds of their prey, building them into traps to capture prey without catching the specific prey's attention.

The imaginative adoption of the prey's sensory world featured centrally in the rodent traps I surveyed in Tanzania. Some, for instance, incorporated enclosed, dark spaces into their designs, mimicking a rodent burrow. Suleimani, a trapper and research technician at the university's pest management center, explained that this is because rodents find wide-open spaces threatening. "They walk next to a wall, or around rocks and bushes, but never across a field," he said. "Panya like small, dark spaces. It is like their home." Individuals like Suleimani draw on a deep well of experiential knowledge about rodent ecology and behavior in fashioning and using traps. Often, Suleimani would share with me behavioral details about panya that are absent from established scientific literature or, occasionally, how they behave in contradictory ways to published reports. On one trapping expedition, Suleimani placed a live trap close to a burrow entrance of a panya buku (*Cricetomys sp.*) and then skipped the next burrow we found. "This is the exit," he said. "Usually, panya buku have territories of around fifty meters, so we have to walk further to set the next trap." Other trappers volunteered behavioral notes when we passed by suitable trapping locations. Once while we were in the mountains, a trapper named Rashidi directed my attention to some long grass. "You will find panya mchanga there," he said, referring to the striped *Rhabdomys pumilio* rodent. He followed this revelation with a description of the scraggy vegetation and lightly disturbed soil that led him to know what species of rodents lived there. Trap technologies are thus imbued with human knowledge *about* animal behavior gained primarily through experience with, and proximity to, a given species. Setting up traps in suitable locations relies on "intimate knowledges" that trappers possess about rodent ecology and ethology.[26]

Taken together, traps materialize processes of knowledge making and imagination that go beyond "the given, the already there, [and] the taken for granted of social life and the world in which social life unfolds."[27] The embodied practices of designing, building, and laying traps are ways that people grapple with the possibilities of living with rodents amid constant struggles to eat and live well.

How to Make a Rodent Trap

Loud squeals of grinding metal competed with traffic noises from the transport stand opposite Iddy's stall. His hoarse voice overcoming the din, Iddy walked me through the steps of making a box trap. He was seated on his goat (mbuzi), the machine that resembled the animal, its neck jutting out to the level of Iddy's face. Attached to the cyborg ungulate's head are bicycle gears, one large and one smaller, conjoined by greasy chains. Iddy had fashioned a kind of handle in place of the pedal, which he turned with one hand to drill through wood held in the other.

Iddy and his siblings grew up in Kilosa village, some seventy miles from Morogoro. "After my mother got pregnant, my father left and I have never received any support from him," he said. In the mornings, Iddy and his elder brother would go and tend to the four-hundred-square-foot field where his mom had planted crops. "My mother would be in town, selling firewood or sugarcane in exchange for maize flour to feed us," he explained. Iddy attributed his difficult life to the fact that he never went to school. "Everything I learn, I learn from the street," he said. He initially worked for food and then later for pocket change (posho) unloading produce from trucks. He roamed the streets and met people who would sometimes offer him construction jobs, such as hauling bricks, cement, and metal.

"It's not always fair [haki]," he admitted, "Sometimes I get paid much less than what was offered, but I never demanded more. I worked hard from morning to night and learned a lot."[28] Iddy's ability to invent new tools like the goat came from having to perform construction tasks without proper equipment. The conditions were often challenging, but he credited those days for gifting him with creativity (ubunifu). "I had to use my brain a lot. My boss was impressed and started paying," he recounted. Eventually, he started saving up wages obtained from his labors. Soon, Iddy was buying Chinese-made traps and selling them on the street.

Moving from town to town, Iddy regularly heard people complain about rodents and other pests destroying their crops. "In one of the villages, I tell you, there must have been something occult [mambo ya ajabu] going on. You could not walk without stepping on a rodent!" he recalled. This gave Iddy the idea of starting a trap business. "But these Chinese traps," he went on, "the customers complain about them." Snap traps imported from China were made of light metal with sensitive triggers. Several customers had returned with complaints that the traps he had sold them maimed rodents but did not kill them. Customers woke up to find blood stains on their sheets and floor, traces of what appeared to be a painful escape. Worse, if a rodent had crawled into a crevice and died, they were often unable to find the decomposing body except by its festering stench.

Iddy realized that he had to make his own traps to accommodate his customers' requests. He experimented with four or five designs, which were all constructed from wood and metal spokes with different trigger mechanisms. Using only hand tools, Iddy created a box trap with a trigger mechanism that he fashioned out of metal spokes twisted into springs. "I discovered that the springs are important. You need enough strength to kill but you don't want it to be too strong," he said, showing me a model that he had just built. "Why not?" I asked. "So that if a child puts her fingers into the trap, she won't get hurt," he answered.

This ethnographic moment reveals how traps are sites for figuring out multispecies well-being. Traps are more than just what Gell called "texts on animal behavior."[29] Rather, as Iddy explained, traps may be designed to constrain and influence the behaviors of rodents and humans who both share a penchant for satisfying their curiosity. For instance, rodents are wary of new objects (neophobic), but they also tend to explore and forage for new sources of food. Similarly, a child's curiosity might be aroused by a trap—a contraption that invites fiddling and play with its dangling bait and mechanical workings. The problem with Chinese traps, Iddy said, is that they are too sensitive. At the slightest touch, they snap and maim, causing the rodent to die slowly and in pain, or in the case of a child, injuring their unwitting fingers.

Iddy's very movements of twisting metal, drilling wood, and fastening a trigger in a trap embody an artisanal calculation that balances the demands of child safety, the need for immediate rodent death, and the efficacy of a trap. "I design traps so the springs work *only* when panya is fully inside and he is killed instantly. The springs are not strong enough to injure a child's finger," he assured me. Iddy does not claim that his traps are humane. However, he respects rodents as living beings capable of experiencing pain, and in some cases, of outsmarting his traps. Some rats, he noted, can avoid getting ensnared. "I haven't found a good design for house rats [panya wa nyumba]. They are too smart [wajanja sana]. They recognize a trap, and very few are tricked," Iddy conceded.

At the market, Iddy's traps are popular because they are cheaper than imported ones and are less likely to fail. Chinese metal traps rust and degrade quickly whereas Iddy's traps, which are made of wood, are more durable in Morogoro's tropical weather. The modular design of Iddy's traps also means that he can easily customize them to specific requests. Fusing business and community interests, Iddy's trap designs draw on his own experiences as a casual laborer to offer better ways to protect people's livelihoods from rodents. As Ron Eglash and Ellen Foster write of maker communities in Africa, Iddy is "simultaneously pulling the warp of innovation geared toward the future while also weaving in the weft of repair practices already deeply entrenched" in their lives.[30] The very practices of drilling holes, bending metal, and hoisting wood into a trap embody the imagination of a future where children are not injured and where rodents are swiftly killed—in other words, where multispecies interests are enfolded into the design of traps.

Rodenticides Are a Poisoned Chalice

Within their political economies of use, traps are material practices that confront us with critical questions about survival, the good life, and multispecies well-being. Traps, trap alternatives such as rodenticides, and deliberations over their respective uses represent the very material ways in which Tanzanians grapple with their own positions within multispecies relations of killing, eating, and living together.

When considering rodenticides, many Morogoro inhabitants are attuned to the risks of toxic exposure. This is reflected in daily conversations about natural products (asili) and locally (kienyeji) grown produce, which they tend to prefer over factory-farmed and store-bought food. "Only foreigners buy frozen store chicken," several Morogoro residents told me, adding, "You don't know what chemicals and antibiotics they pump into them." People also tend to buy produce on the street or in the wet markets, sold by women "from the mountains" that are "free from pesticide." Stacey Langwick noted that Tanzanian gardeners harbor similar suspicions toward industrially produced food. Practices of cultivating medicinal foods (dawa lishe), Langwick writes, are sites of meditation and mediation for cultivating a politics of habitability amid an industrializing Tanzania.[31] For the same reasons, small-scale Tanzanian farmers with whom I spoke rarely use rat poison (sumu). "We don't know if these chemicals go into our food or our water," they mused.

Shawa is a retired agricultural officer at the Rodent Control Centre located along the main road to the Sokoine University of Agriculture. He commands the respect of all current staff and still regularly comes by the office. During the early days of the centre in the late 1980s, Shawa conducted several studies monitoring long-term population fluctuations of panya shamba, or field mice (*Mastomys natalensis*). He performed several "palatability studies" in which he tested several mixtures of bait with poison to see which ones attracted (and killed) the most rodents. In our interview, Shawa explained that the Rodent Control Centre was initially established to improve Tanzania's agricultural sector by providing advice and technical assistance to farmers dealing with rodent pests. More recently, however, the under-resourced centre functions as a clearing house for government-distributed rodenticides during outbreaks. These included several varieties of warfarin and zinc phosphide, well-known poisons used throughout the world to combat rodent infestations. Shawa worried about the long-term health effects of these poisons. "I see that people who used poisons in the 1990s, they now have some kind of illness," he explained. When I asked him to elaborate, he recalled that farmers developed growths on their hands and had difficulty clenching their fists.

Shawa was skeptical when I told him that I could not find any published research on the long-term effects of warfarin on the health of humans, other animals, or plants.[32] Most of the studies dealt only with measures to prevent the accidental, immediate poisoning of livestock and people. "The problem with using poisons," Shawa noted with concern, "is that you use a lot and so you have huge sacks of it lying around that farmers didn't use, even today. Who knows what happens to the poisons? Are they seeping into the ground? Are they going into the well water?" Shawa continued, "Children may die because they eat rodenticide. Disposing of poisons is a challenge." He recounted how rodents could build resistance to these rodenticides so that farmers must use second generation versions to keep up. Shawa opined that the centre's main job should be to educate farmers about farming responsibly, including the responsible use of poisons. Yet, with the centre's reduced budget, deteriorating equipment, and dwindling staff, this was difficult to do. Common to many scientific institutions throughout Africa, these challenges cause Shawa to worry over the fact that he was never able to study the unintended, toxicological consequences of rodenticide use.[33]

Iddy echoed Shawa's sentiments about poisons: "I tell you, some of these poisons take time to work, up to seven days. By then, the rodent would have gone far. What if someone eats him? What if a cat eats him? Where does the poison go?" Like Shawa, Iddy questioned what happens to poisons once they have been ingested by the rodent. "Do they end up in the water, in our food? When you poison a rodent, you poison other animals together," he said. For this reason, Iddy discourages his customers from using rodenticides. His advice is always to use a trap, or raise a cat, but never to resort to rodenticides. "Our body might transform when we eat something that has rat poison," Iddy conjectured. "And the rodents suffer. They don't die right away. They crawl around, they go mad, they go to a corner, and then they slowly die."

Iddy's distrust of rodenticides articulates a particular stance in relation to multispecies justice. Iddy arrives at his position through his work of inventing and building traps as he generates alternatives to rodenticides that nonetheless remain imbricated with important questions about human-rodent relations in an agricultural context. In this regard, trap making is both a practical and a theoretical endeavor. Through the very handiwork of building traps, Iddy thinks about and imagines a future that is just possible for the intertwined lives of people and rodents in Morogoro.

Toward a Generative, Multispecies Just Possible

Iddy had been invited to set up his stall at the Annual Nanenane Agricultural Fair, where I sought to meet him. In a long queue to enter the fairgrounds, people shuffled slowly under the midday heat. I could smell the charred fat of mishikaki (skewered meat) and roasting popcorn. Buses brimmed with school children in bright white uniforms, jostling for space and prompting piercing shrieks from the policewomen and their whistles.

Once I finally bought a ticket and entered the fairgrounds, I waded through the crowd to Iddy's stall. On the way, I passed by shiny tractors, a patch of gigantic eggplants, a snake gallery, and a mock Bwana Sukari factory demonstrating how sugar is made. At last, I found Iddy at his stall, seated on his goat. I was surprised to see that he was drilling through a piece of metal rather than the usual wood. This was a new design. He had also repainted the goat in bright colors and added a new motto: "Tanzania ya Viwanda. Morogoro Kwanza. Ubunifu jadi yetu." In English, this translates as "Industrial Tanzania. Morogoro Comes First. Imagination is our heritage." Iddy's stall was shot through with the country's flag colors and symbols, thus positioning his work as part of a national aspiration that boasts creativity as traditionally Tanzanian.

As I watched Iddy work, I soon recognized his new trap design. It was a Sherman. Made of aluminum and light to carry, these live traps are the tool of choice for ecologists conducting trap-and-release studies of small mammal populations. And just a few days earlier, over a hundred Sherman traps laid out overnight by Sokoine University's Pest Management Centre had been stolen. Data collected from this study was intended to contribute to a long-term project to predict rodent outbreaks and implement pest management strategies that did not rely on poisons. The trappers Suleimani and Rashidi had been able to track down and retrieve several stolen traps at Chamwino market. Yet, they only recovered several dozen, and the research had to be halted.

This was where Iddy came in. He had bought sheets of aluminum, cut them into smaller pieces, and constructed several Sherman-like traps. "I'm still testing the trigger springs," he remarked. He inserted a pencil into the trap, which meekly snapped shut. Thanks to Iddy's ability to reverse engineer a Sherman and construct the trap with an entirely different spring mechanism, the university research project was able to continue. The university could buy the traps more cheaply, without incurring the exorbitant import duties that Tanzanian customs frequently levy. With Iddy's technical ability, Suleimani and Rashidi continued to trap mice, and their data collection was only briefly interrupted.

Both Iddy's kill traps and live traps are interventions into an ongoing multispecies predicament that binds humans and rodents into close-knit relations that require constant negotiations of who eats, who dies, and who lives. Val Plumwood orthographically recognizes this intimate relation as "Food/Death," writing that the most "basic feature of animal existence on planet earth" is that "we are food and that through death we nourish others."[34] In Morogoro, where people rely on food they cultivate to make ends meet, human-rodent entanglements become sites where nourishment and death must be constantly negotiated. Rodents who consume too much food threaten human sustenance and endanger lives that depend on making just enough.

Trap making is first and foremost Iddy's means of earning a living. He is proud of his accomplishments, particularly given his journey from the days of moving hundred-pound loads in exchange for food. By serving his community's needs for rodent control, generating income, and eschewing the accumulation of profits exclusive to an owner of intellectual property, one could argue that Iddy's audacious creativity also proposes new ecological entanglements that try to resolve the Food/Death conundrum. Whether they are designed to kill quickly and thoroughly, or for live trapping so as to offer safer, poison-free methods, Iddy's traps knit together—materially and intellectually—human and rodent worlds. His traps make visible the potentially disastrous, cascading ecological consequences that ensue when rodenticides are used, in hopes of avoiding what Deborah Rose Bird calls "double death."[35] The fact that people may consume rodents and other plants and animals that have been exposed to toxic rodenticides means that the use of any rodenticide runs the risk of jeopardizing many lives. From the perspectives of Iddy, Shawa, and others who grow their own food, the use of rodenticides conjures anxiety about wide-ranging, long-term effects of poison on human and environmental health. Additionally, "double death" conjoins shared, multispecies vulnerabilities: rodents and children getting maimed by badly designed, faulty traps imported from abroad.

Tanzanians who use traps do not deny that traps mark the end of an animal's life. Yet, despite the many methods for trapping and killing rodents practiced in Morogoro, Iddy and others readily concede that their machinations may be foiled by "smart" (wanaoakili) rodents. Even when faced with alluring (albeit poison-laced) baits, rodents adapt over the course of a few generations and build resistance to rodenticides. Rodents' ability to survive and subvert the most enticing of traps garners Iddy's admiration. "In the end, you can only do so much. Rodents are cunning," he concluded. "You can lay a trap but they know, and they will go around it and eat your maize."

What is valuable, then, for people in Morogoro living with rodents is not the total eradication of rodent pests through indiscriminate methods such as rodenticides. Rather, value is generated in the everyday endeavor, through the design and deployment of technology, to live well with those who eat together. These endeavors take on a concrete form in the traps that Iddy makes. When deployed, these traps become significant sites for reconfiguring relationships between rodents and people, informed by a constantly negotiated calculus of multispecies nourishment.

Conclusion: Imagining the Just Possible

Such a history begs the question, How does one delight in precarious life?
—**Joshua Bennett**, *Being Property Once Myself*, 8

Not so long ago, the white-minority governments of Rhodesia (present day Zimbabwe) and South Africa used warfarin and other rodenticides as chemical weapons against Black activists fighting for decolonization.[36] White supremacists in southern Africa saw little difference between Africans and rodents, and they sought to eradicate both. Black Americans, too, have been dehumanized by racist violence and other experiences of inequality that often placed them in close disposition and proximity to nonhuman animals, including rats.[37]

Against the backdrop of these histories, Joshua Bennett counters that such dehumanizing experiences prepare the ground for Black writers to articulate a "more robust vision of human, and nonhuman," and its "cognitive and otherwise potential."[38] For Bennett, it is important that his work acknowledges Black experiences of suffering and subjection without foreclosing possibilities for poetry, imagination, and resilience.[39] The stories I tell here of trap making in Morogoro attempt to answer Bennett's question, "How does one delight in precarious life?" Although Iddy and others who work with rodents in Morogoro live on the edges of making ends meet, they find delight and formulate visions of the future through their design and deployment of traps. It is for this reason that Iddy's trap making is a form of generative justice.

By designing traps that subvert their commodified counterparts, Iddy generates new spaces within which he and others in Morogoro can reconceive their social relations with one another and with other nonhuman animals.

In contemporary Tanzania, human-rodent relations manifest the practical realities of learning to live well with others—a theme of central importance to multispecies justice. People who opt to use Iddy's traps seldom appropriate the language of war against rodents that is characteristic of pest extermination efforts in Euro-America. Instead, they embrace them as part and parcel of everyday life. "They live with us, they eat with us," a fruit seller at the market once told me nonchalantly. While many of my interlocutors have relied on terms like "enemy" (adui) to denote panya, rarely did they want to see them eradicated or killed by the thousands. If anything, rodents were acknowledged for their intelligence and resilience, even if begrudgingly. "If only humans [binadamu] were more like panya!" said Rashidi, in the context of deploring the so-called antics of today's youth. For Rashidi, many young Tanzanians dress sloppily and have abandoned all effort to look presentable, behavior that paled in comparison with the conscientious, self-grooming habits of rodents. Human-rodent relations in Morogoro thus exemplify a cosmopolitics wherein possible notions of justice are not foreclosed by a particular view of rodents but rather worked out in the design and use of traps.[40] This cosmopolitical approach draws attention to the material ways through which people conceive of and enact justice, and how these practices relate in turn to the access and distribution of resources and technology.

The (unequal) material conditions that undergird Iddy's trap making came through in our final conversation. When asked about his hopes for the future, Iddy laughed. He gestured to his traps and said:

> First, I would like a power drill. A drill will let me make four times more traps. Second, I would like a factory. I want to provide jobs for youth who cannot find any work; if you don't have work, you don't have nothing [bila kazi, hamna kitu]! Third, I would like some stickers to put on each trap, with my name and phone number, so people know that this young man from Tanzania made this trap. It is the only one like it in the world, and when people in China, Malaysia, America see the trap, they know that this man from Tanzania, who never went to school, made this trap.

His technoscientific dreams notwithstanding, Iddy's desire to own an electric drill should caution us against celebrating this story merely as an example of African improvisation or a smart work-around. Iddy would not have chosen to make traps using the goat if he could have done otherwise. It is for this reason that I have avoided using the terms *improvisation* or *bricolage* to describe Iddy's traps because these terms have so often and subtly marked African practices of technology as inferior copies of those found elsewhere.[41]

Instead, I appropriate the language used by scholars of computing, who credit hackers and makers for their ingenuity in designing open-source software that reconfigures existing notions of equality, freedom, and justice. By smuggling notions of activist creativity into Iddy's trap making enterprise, I avoid framing African hacks into extant technologies as improvisation, which as the term's etymology suggests, describes an unforeseen progress. On the contrary, Iddy's traps are purposefully designed through an intricate, intellectual process that brings together questions of livelihoods, well-being, and multispecies justice. They are ubunifu, or inventions, which at their roots in both English and Kiswahili foreground the new and deliberate, both as idea and object. And they arouse feelings of delight and pride in Iddy, who continues to show them off to passersby and potential customers.

Seriously engaging with both human-rodent relations and with the hardships and possibilities posed by such relations forms the ground upon which people like Iddy envision just-possible futures. It is within such knotty multispecies relations that Iddy finds delight and pride—so colorfully conveyed on his machine—in showcasing a vision for his trap enterprise and for the world. To be sure, Iddy's vision of life can be stark. "Maisha ni mapambano," he often says, "Life is a struggle." Although Iddy works under challenging circumstances and within limited resources, earning just enough money to get by, his traps are nonetheless modes of self-expression, pride, and aspiration. By "imagining other possibles and other realities" through his trap designs, Iddy, to borrow Arturo Escobar's words, "forces us to rethink many of our everyday practices and politics."[42] Seated on his goat, turning the drill, and constructing traps, Iddy crafts *just*-possible futures, in which he would own a trap factory that provided jobs to his community while his traps circulated across the world.

The just possible, as Iddy's story suggests, is the condition of doing enough to thrive while incorporating considerations of more-than-human well-being with ingenuity. It is a condition that acknowledges the radical potential in particular and local practices of kufanyafanya tu, or "making do with what one has."[43] Even as they evoke elegant objects of contemporary art and contemplation (see figure 7.3), as Alfred Gell would have appreciated, Iddy's traps embody his ubunifu (imagination) for crafting just possibles. They are informed by a striving to live well and delight in multispecies

7.3 One of Iddy's traps. Photograph by Jia Hui Lee.

worlds. Meeting the needs of Tanzanian farmers, whose livelihoods depend on safeguarding sufficient harvests from rodents, Iddy generates designs for killing *and* living with rodents without indiscriminately endangering the people, plants, and other animals who share the agricultural communities of Morogoro.

Notes

1. All names except Iddy's are pseudonyms.
2. *Mboga* literally means vegetables, but the word is also used to denote small, edible creatures including mice and termites. In Tanzania and throughout southern Africa, rodents are occasionally trapped or hunted as food. A rodent trainer from Iringa told me that rodents are considered a meat relish, or a "bonus addition" to the main meal, or *kitoweo*.
3. See Brooks, "Animal Rights and Vertebrate Pest Control."
4. Littin et al., "Humane Control of Vertebrate Pests."
5. John Hadidian notes that many of the terms used by advocates and critics of animal welfare such as *pest* and *humane* are not clearly defined. Hadidian particularly points out that the use of so-called humane traps that restrain or capture animals alive may often result in lacerations, trauma, and even death when an animal is left out in extreme heat or cold. What animals are seen as pests can often change depending on particular ecologies, communities, and histories of migration and colonialism. See Hadidian, "Taking the 'Pest' Out of Pest Control"; Hadidian, Unti, and Griffin, "Measuring Humaneness." I thank Kat Poje for pointing me to Hadidian's works.
6. Mbembe, "Necropolitics," 11.
7. I am indebted to Clapperton Chakanetsa Mavhunga and Laura Ann Twagira's work for broadening the space to think about technological innovation from the continent. Mueni wa Muiu and Guy Martin use fundi (fabricator) as an analytical concept in political science. Kenda Mutongi encouraged me to think deeply about Iddy's ingenuity and his relationship to working on the streets of Morogoro. See Mavhunga, *Mobile Workshop*; Twagira, "Introduction"; Muiu and Martin, *New Paradigm of the African State*; Mutongi, *Matatu*, 271.
8. Eglash, "Introduction to Generative Justice."
9. Iris Young critiques the focus on distribution in social justice movements that pits different social groups against one another. She advocates for an "enablement" approach to justice, which emphasizes eradicating "structural injustices" that affect some groups more than others. In this respect, her work is similar to the capabilities approach of justice later developed by Martha Nussbaum and Amartya Sen. See Young, *Justice and the Politics of Difference*; Nussbaum and Sen, *Quality of Life*.
10. Gell, "Vogel's Net," 26.
11. Morgan, *American Beaver and His Works*, 236.
12. Alfred Gell wrote his essay in 1996 whereas Lewis Henry Morgan's book on *The American Beaver and His Works* was published in 1868.
13. Haraway, *When Species Meet*, 295.
14. Quoted in Mura, "Discontented Farmer."
15. National Bureau of Statistics, United Republic of Tanzania, "Gross Domestic Product 2017."
16. Haraway, *When Species Meet*, 80.
17. Mavhunga, *What Do Science, Technology, and Innovation Mean from Africa?*, 8.

18 Kirksey, *Emergent Ecologies*.
19 Jacques Derrida coined the term *other others* to refer to nonhuman animals who often fall out of human-centered ethical considerations: Derrida, *Gift of Death*, 69. I thank Sophie Chao for bringing this to my attention.
20 See also Jiménez and Nahum-Claudel, "Anthropology of Traps."
21 Tylor, *Anthropology*.
22 Lips, *Origin of Things*, 83.
23 Lips, *Origin of Things*, 80.
24 Gell, "Vogel's Net," 29.
25 Gell, "Vogel's Net," 29.
26 Both Hugh Raffles and Radhika Govindrajan write about intimacy and knowledge production in multispecies relations. Even Lewis Henry Morgan, in his work on a different, larger rodent, acknowledged the "knowledge of the habits of beavers [that] is necessary to the trapper to pursue his vocation." These trappers were "Indian and white trappers on the south shore of Lake Superior." See Morgan, *American Beaver and His Works*, 227, 133. See also Raffles, "Intimate Knowledge"; Govindrajan, *Animal Intimacies*.
27 Joel Robbins makes a case for anthropologically studying people's imaginations of alternatives and possibilities. Robbins, "Beyond the Suffering Subject," 457.
28 *Haki*, the Kiswahili word translated as "fairness" or "justice," is also the word used for rights. *Haki za binadamu*, for example, means "human rights." Haki is one of those Indian Oceanic words that shaped and marked Tanzanian cultural practice. I have encountered haki in the context of justice and fairness in Bahasa Melayu, Hindi, Urdu, and Arabic. See Geertz, *Local Knowledge*, chap. 8, for ethnographic examples of haqq from Indonesia and Morocco.
29 Gell, "Vogel's Net," 27.
30 Eglash and Foster, "On the Politics of Generative Justice," 129.
31 Langwick, "Politics of Habitability."
32 The World Health Organization (WHO), United Nations Environment Program (UNEP), and International Labor Organization (ILO) of the United Nations concluded, based on available studies, that "exposure of the general population to warfarin as a rodenticide through air, drinking-water, or food is unlikely and does not constitute a significant health hazard." International Programme on Chemical Safety, "Warfarin," https://inchem.org/documents/hsg/hsg/hsg096.htm#-SectionNumber:2.7. Gwen Ottinger and others term this lack of research on a chemical's long-term health hazard as a "structured knowledge gap," meant to disempower communities and exclude them from procedural justice. See Ottinger, "Changing Knowledge." It is a webpage with text so there is no page number.
33 See Tousignant, *Edges of Exposure*.
34 Plumwood, "Tasteless," 324.
35 For Deborah Bird Rose, the death of an organism, ecosystem, or metabolic pathway usually results in a "relentless cascade" of more deaths, "fracturing a compact [beween life and death] that has been integral to life on earth." "Double Death."
36 See Gould and Folb, "Project Coast"; and Wittenberg, "Poison in the Rhodesian Bush War."

37 See Mavhunga, "Vermin Beings."
38 Bennett, *Being Property Once Myself*, 8.
39 Bennett, *Being Property Once Myself*, 8–10.
40 Isabelle Stengers outlines a deliberative framework for envisioning a world we want to live in that considers the experiences and existence of different actors—human and nonhuman—without foreclosing the political possibilities that emerge. Trap making could be considered a cosmopolitical practice according to Stengers's work. See Stengers, "Cosmopolitical Proposal."
41 Lily Irani critiques the term *jugaad* (work-around), as used by Indian entrepreneurs to describe rural technologies. Calling a technology jugaad ascribes it a lack of design, inferior to proper innovation. See Irani, *Chasing Innovation*, 175–92. Chakanetsa Mavhunga writes that "tinkering" is "such a horrible word because it refers to a mender ..., a trial and error person, a meddler, or, worse yet, a clumsy, unskilled worker." *What Do Science, Technology, and Innovation Mean from Africa?*, 7–9. I am also grateful to Jean Comaroff for helping me think through these points in a conversation about Bedford lorries in Sudan.
42 See Escobar, *Pluriversal Politics*, 4.
43 Mutongi, *Matatu*, 35.

8

Inscribing the Interspecies Gap

M. L. Clark

8.1 (*previous page*) Original drawing by Feifei Zhou.

But the bridges were broken between him and me, because what was his obsession is now an axolotl, alien to his human life.

—Julio Cortázar, "Axolotl"

WRITING ON JUSTICE is an act of routine compromise. Whether in fiction, formal essays, or the rallying cries of direct activist discourse, we draw upon the power of familiar terminology to establish hierarchies of concern, then hope that our intentions with far-reaching and nuanced concepts like restorative and rehabilitative justice can span the experiential divide. In the multispecies realm, though, this divide is unbridgeable with text alone. And so, the compromise broadens in scope. Writers on justice in a speculative mode become self-appointed speakers for those who cannot converse with us on the page. How can we address the linguistic insularity of all attempts to write our way into better worlds?

Many of my speculative stories plainly address the limits of our deliberations on justice. In my first published piece, "Saying the Names," a scientist attempts to "speak" a member of an alien species (one in which language drives biological transformation) into physically reconfiguring into a form more familiar to him. This attempt ends in a death. The story follows the scientist's daughter as she seeks to understand what happened and why, so as to defend her father in the ensuing murder trial. The story's dominant human languages—legal and scientific—stand in pointed contrast to the daughter's realization that even if her father did not murder the alien by formal standards, an injustice nevertheless took place.

But with what vocabulary might one even begin to speak such truths?

Subsequent tales, such as "Hydroponics 101," "The Stars, Their Faces Uplifted in Song," "Seven Ways of Looking at the Sun-Worshippers of Yul-Katan," "Belly Up," and "A Tower for the Coming World" illustrate a similar focus on the limits of justice, though these stories are centrally concerned with justice in humanoid realms. "Hydroponics 101," for instance, presents a prisoner isolated with a substance that can provide for all his needs if he learns to live in harmony with it, by empathizing with something external to himself. But the greater challenge lies in his return to a world that may not be ready to accept rehabilitative justice: to empathize with him in turn. All these stories ask: Can individual transgressors ever be healed, and individual transgressions surmounted, within massive enterprises of ongoing oppression? How do we approach marginalized cultures when their internal traditions of oppression ultimately serve the interests of wider systems of power?

Even when dwelling only on the pursuit of better justices for fellow human beings, our literary vocabularies are heavily compromised. Mainstream Western discourse now readily accommodates seasoned users of formal terms like *colonialism* and *the Other*, which does not bode well for the actual progressive quality of related discourse. If it is easy within a system of oppression to speak of specific demographics as oppressed, to articulate the sites of their oppression, and to cast ourselves as speakers for their plight—all without meaningfully enacting material changes—then perhaps the core injustices of our world run deeper than any mere tinkering with specific formal terms can fix.

This is the thought, at least, that informs the work of many speculative writers from non-Western contexts. Nnedi Okorafor, author of the acclaimed Hugo- and Nebula-Award-winning *Binti* series, expressly refuses the term *Afrofuturism* for her work, because her Nigerian speculative fiction is not intended to serve Western projects of postcolonialism or related racialized-justice resistance from within that highly adaptive and all-consuming cultural paradigm. Rather, she is a writer of Africanfuturism and Africanjujuism: two terms that she asserts as belonging to an African literary discourse that is subordinate to no other.[1]

Nor is she by any means the first thinker to repudiate the notion that Western histories of injustice can be solved with Western depictions of the problem, let alone paths to utopia. Frantz Fanon warned those who would seek to define colonialism—a major and ceaseless project even fifty years on—that "it is utopian to try to ascertain in what ways one kind of inhuman behavior differs from another kind of inhuman behavior."[2] Ursula K. Le Guin, following from Fyodor Dostoevsky's "injustice of Heaven" discourse in *The*

Brothers Karamazov (1880), likewise articulated distrust in utopic projects as invariably arising from unjust social contexts. In her speculative parable, "The Ones Who Walk Away from Omelas" (1973), the idea of paradise becomes unconscionable when even one must suffer so that the rest may thrive.

If we struggle even to find a common language for transformational justices among fellow humans, what hope have we of finding the words to speak of restoration and rehabilitation as it relates to multispecies trespass, too? In "Axolotl," Julian Cortázar advances the idea that any human obsessed with understanding other species might only create new schisms in the process. This fabulist tale suggests that one might devote one's life so immersively to the study of another species that one ends up a part of it, on the other side of the glass, but even so complete a personal understanding would fail to shatter the barrier between worlds itself. Is there any more honest way of depicting human aspirations toward multispecies harmony?

I broached this notion of limitations to human understanding in "To Catch All Sorts of Flying Things," another science-fictional "detecting" tale in which our human protagonist uncovers the truth behind a death on an alien world. Greysl, the human security liaison for the Partnership colony, learns not only who destroyed the last egg of an alien species but also how the use of human language (to establish the nature of this crime and to shape investigation around it) hindered a swifter meting out of justice in the most restorative form possible. Once the hows and whys of this genocidal tragedy emerge, our human protagonist is also not the primary agent of ecosocial restoration: Greysl can only bear witness to how the culpable party, from a different species, decides to provide restitution for the dead.

My subversion of a common storytelling structure, in which humans are configured as centrally important to the naming of multispecies injustices and the pursuit of their resolution, was made possible because the world of this story, Drasti Prime, is chockfull of discourses unfolding outside human hands: outside, even, most human notions of discourse itself. The world was shaped by an ancient, more advanced species that no longer resides within the region. These Makers left behind Spinners—giant arachnoid biocomputers just on the cusp of surmounting their original programming, who make their most important decisions out of human sight—and Chiggers, a semifantastical reimagining of nanobots as keepers of preconceived notions of planetary balance, which they achieve by infecting foreign entities with blueprints that trigger transformation into native species.

Then there are the other, nonhuman species that the Spinners, led in this story by a representative called Kurrick, permit to set up colonies on the planet's surface: the Feru, a fungoid species with a different approach to intimacy and interconnectivity; the Esh, a nebulous cloud that struggles with questions of autonomy and collective justice over the course of its dispersion/reintegration lifecycle; the Saludons, a military compact of conquered peoples that have bonded with one another through outsized pride for having been assimilated into a larger identity and purpose; and the (late) Obys, a spacefaring species that was already dying off from parasitic infection before the act of genocide.

Greysl, who struggles to maintain healthy relationships even with just two other humans in the Partnership colony, tries to narrate dominance over this terrain and its mysteries in the course of the story's investigation—and fails. However, the end of "Flying Things" is hopeful, despite the revealed inadequacy of all efforts to impose human conceptualizations of trespass and justice onto other species. Greysl, for instance, comes away from the experience with key insights that invite a renewed focus on improving fraught, intimate relationships (in this case with two lovers) within the human realm. Meanwhile, Kurrick and her fellow Spinners enact what they regard as an appropriate restorative act, by allowing their bodies to be transformed into members of the extinct species. But an even more vital lesson emerges in the last exchange between these two, when Kurrick explains to Gresyl the choice that the Spinners made:

> "We are—out of balance," she had explained instead. "As such, members of our own account will provide the necessary adjustment."
>
> "But—you'll lose yourselves in the process," I'd countered. "You'll become Obys completely. One life sacrificed for another—that can't possibly be a fair accounting either."
>
> Tic-tic-tic-tic. The last of those damned spooky calculations I'd ever hear her make.
>
> "It is—no less than The Makers did, when they too erred in their accounting."
>
> And with that ... well, whew, talk about your offhand remarks containing multitudes. But even without pressing further on this accord—even without accidentally losing myself again to my species' relentless need to impose narratives on everyone around them—I understood at once the gist of what she must have meant [...] A whole other species lost to some ancient error on The Makers' part—either of a moral nature or purely accidental. A whole other slice of alien history on Drasti Prime.[3]

Greysl never learns this "other slice of alien history." But by the story's end, our self-appointed investigator of crimes defined by human terms has at least gained a healthier respect for the limits of individual knowledge and come to appreciate that not every multispecies trauma is, will be, or perhaps even should be accessible to prying officers from the Partnership.

In the realm of my published work, I have already returned twice to the universe of this Partnership: once, on Drasti Prime, to explore the Feru's more immersive forms of intimacy in contrast to a toxic form of human relationship; and again, on a nearby station, to invite readers to see more of the Partnership's linguistic oppressions from the eyes of an outside, alien species. I will no doubt return to this universe in future published stories, too, because its narrative parameters and world-building have provided this writer of speculative fictions with a means not only of talking about a wide range of intra- and multispecies justices but also of foregrounding the inherent concessions in any attempt to speak on such themes.

Does one need an expressly speculative and fictional mode to speak frankly about the ways in which our discourses sometimes reify the very injustices we seek to name, confront, and overcome? Hardly. But whatever literary style one might use to advance their own exploration of the problems of past, present, and future trespass, let the writer on such themes of justice remember the extent to which they are always making choices not just to speak for others but also to speak over others: over forms of multispecies discourse, that is, which exist on no page of written text; and over the intrinsic gaps between every living being's distinct experiences of consciousness, trauma, and trespass.

Any pursuit of multispecies justice that elides the existence of this gap is at risk of becoming colonization by some newer, more self-serving name.

What better forms of justice await us, in learning to inscribe the gap itself?

Notes

1 Okorafor, "Africanfuturism Defined."
2 Fanon, *Black Skin, White Masks*, 63.
3 Clark, "To Catch All Sorts of Flying Things."

9

Nuclear Waste and Relational Accountability in Indian Country

Noriko Ishiyama and Kim TallBear

9.1 (*previous page*) Original drawing by Feifei Zhou.

> We would like all geographers, all people, to learn what they can from us, about what it might mean to live in a world that is relational, that co-becomes with us and each other, that is knowing, that is alive—even in its death.
> —Bawaka Country et al., "Co-becoming Bawaka," 456.

NUCLEAR WEAPONS AND ENERGY PROGRAMS in the US American West have produced land-use conflicts that are situated within a history and geography dominated by settler-colonial militarism. The US West, almost half of which is still under federal ownership and control, has served a wide range of nuclear projects since the 1940s. Ecological destruction generated by nuclear development in the service of US-settler national security goes hand in hand with the systematic elimination of Indigenous Peoples and their lifeways. These Peoples and lifeways have been for millennia coconstituted with the same lands. In order to survive such violent colonization and maintain political and cultural sovereignty,[1] tribes in lands now occupied by the United States work to maintain constitutive relations with and responsibilities to multispecies and other more-than-human relations within their traditional land bases.

This chapter brings together geographical and environmental-justice analyses of land use with Indigenous studies' legal and relational approaches to examine land-human relations as they are constituted in relationship to two different landscapes. The first landscape discussed in this chapter surrounds the Hanford Site in Washington state—a site appropriated and reshaped for the purpose of plutonium production under the Manhattan Project during World War II. The second landscape in the state of Utah is what many people assume to be a desolate desert called Skull Valley, where in the 1990s a small tribe attempted to site an interim storage facility for high-level radioactive waste. Both of these places are located in what US mythological narratives might call the Wild American West. Both sites play important roles in the history of nuclear development.

As part of clarifying "the violent disruption of human relationships to the environment," in the words of Potawatomi philosopher Kyle Whyte, we examine interdependent relations between Indigenous Peoples who call these landscapes home and their more-than-human relations in these places, including both material and "spiritual."[2] Heather Davis and Zoe Todd, in their analysis of the so-called Anthropocene, explain how such physical violence to Indigenous Peoples and their more-than-human relations is simultaneously conceptual violence and erasure of specific Indigenous practices, environmental relations, and places:

> The Anthropocene as the extension of colonial logic systematically erases difference, by way of genocide and forced integration and through projects of climate change that imply the radical transformation of the biosphere. Universalist ideas and ideals are embedded in the colonial project as it was enacted through a brutal system of imposing "the right" way of living. In actively shaping the territories where colonizers invaded, they refused to see what was in front of them; instead forcing a landscape, climate, flora, and fauna into an idealized version of the world modelled on sameness and replication of the homeland.[3]

Contrary to the universalizing and colonial project of nature versus culture — and the similar binary response by some settlers in their articulation of the Anthropocene — we approach human-place relations via a framework that privileges specific co-becomings of lands and water with First Peoples in specific places.[4] Co-becomings are entangled biological and social relations that can be understood in part via contemporary social and natural science approaches to apprehending the world. These co-becomings can also be understood in part via core Indigenous historical narratives and stories for living well together.

The idea of co-becoming that refuses the nature-culture divide is also related to our soft refusal of the word *spiritual* which — while in common use among tribes in the United States and other Indigenous Peoples — is also burdened in both non-Indigenous theological and philosophical meanings by a separation of "matter from the immaterial, and thus the knowable from the illogical."[5] In order to unburden our analysis from those binaries, we deploy the notion of being "related" (which Shorter recommends and which is ascendant in critical Indigenous studies globally) more than we use the term *spiritual*. While the Indigenous communities with whom we learn do often use the language of spirituality, our experience is that materiality—the essential role of land and material existence—is conjoined with the immaterial or the unknowable when Indigenous speakers invoke spirituality. But this complex idea may not be adequately understood by especially non-Indigenous readers. "Related," on the other hand, in Shorter's words "emphasizes mutual connectivity, shared responsibility, and interdependent well-being."[6] Consequently, instead of presenting a romanticized, spiritual, and overgeneralized notion of Indigenous environmentalism, we also emphasize how specific, dynamic Peoplehoods are coconstituted materially and socially with ecological processes produced in particular historical and geographical contexts.

In addition, we tend not to use multispecies terminology, but thinkers working within that set of terms and citations will find resonance with our analytical frameworks drawn primarily from Indigenous studies and geography. We also draw on Native legal frameworks to help us demonstrate Indigenous struggles for self-determination or sovereignty in our analysis of those particular tribes' struggles for cultural authority and contemporary legal jurisdiction over their traditional lands.

What follows results from twenty-five years of conversation related to our respective research programs. Noriko Ishiyama is a geographer from Tokyo, Japan, and Kim TallBear is an Indigenous studies scholar and anthropologist of settler-colonial science who hails from US-occupied Dakota homelands (Sisseton-Wahpeton Oyate) and now resides in Treaty 6 territory (today known as Edmonton in Alberta, Canada), a traditional gathering place for diverse Indigenous Peoples including the Cree, Blackfoot, Métis, Nakota Sioux, Iroquois, Dene, Ojibway/Saulteaux/Anishinaabe, Inuit, and others. The cases we highlight demonstrate legally and culturally complex social processes—both Indigenous and settler-colonial processes—that frame Indigenous People's struggle for self-determination.

When we speak of *self-determination* or *sovereignty*, we use those terms keeping in mind Native American legalese that is sometimes contested as too colonial but which helps tribes in limited ways defend land and jurisdictional rights. And we keep in mind broader cultural meanings such as Wallace Coffey and Rebecca Tsosie's notion of "cultural sovereignty," "that is, the effort of Indian nations and Indian people to exercise their own norms and values in structuring their collective futures."⁷ We do the best that we can in English and in being so viciously constrained for centuries by settler-colonial thought.

Like spirituality, sovereignty has conceptual and ethical limitations that lead us again to privilege relational frameworks also privileged in Indigenous communities. Relational frameworks in part critique the limitations of settler-colonial concepts such as sovereignty predicated on individualism, rights over responsibilities, and the nature-culture divide that renders relations—both human and more-than-human—into property. You will see that we refer to the struggles of tribes we have studied and worked with (TallBear also was an environmental policy specialist prior to becoming an academic) not only as struggles for land rights but also as struggles to recover relations after settler-colonial assaults on the very existence of the tribes in question. The tribes whose struggles are documented in this chapter seek to fulfill both human and more-than-human responsibilities—their responsibilities to care for the land and to survive as distinct Peoples coconstituted with those lands.

You may notice the subtle absence of the word *identity* in this chapter. We have debated calling attention to it, but it seems important to highlight why we choose not to use identity so the reader can better apprehend what relations gets us in its stead. Identity is a concept that is integral to the Eurocentric colonial assault on human and more-than-human relations on these continents. As a popular concept, it does not necessitate ongoing relating. For example, identity is often used to refer to discrete biological conjoinings within one's genetic ancestry. Discovering shared genetic markers with another individual (say a biological ancestor) or an ethnic population can in theory spur social contact, but such biological connections are often fetishized. This means they are used to stand in place of actual social relations, to support an individual sense of self via claims about who one *is*. Identity can undercut notions of *becoming in relation* because biological traits and connections become properties and property. They need not be lived socially. Indeed, as TallBear has written elsewhere, settler relations are property.⁸ In this chapter, we refuse identity and its property concepts when possible and substitute relational concepts that encompass both material (sometimes biological) and social relations between humans and also with more than humans. While we must sometimes work with the property concepts of the settler state and its legal infrastructures, we also highlight when possible a relational framework to emphasize co-becoming, which can lead to better relations in the world.

Environmental History of the Hanford Site

The Hanford Site occupies 584 square miles on the Columbia River in southeastern Washington state in the Pacific Northwest of the United States. It is located north of the confluence of the Yakima, Snake, and Columbia Rivers that have supported the livelihoods of multiple Indigenous communities for centuries. While the Wanapum had their land base within present Hanford boundaries, neighboring tribes—including the Yakama, Nez Perce, Cayuse, Umatilla, and the Walla Walla—also hunted, gathered, camped, prayed, and engaged in other social relations with other Peoples and also with multispecies communities in this area.

Before Euro-American settlement, Indigenous Peoples in the area developed a deep understanding of the local ecology by nurturing interdependent relationships with one another and with the landscape. Negotiations for the treaties of 1855 established the Yakama Indian Reservation, now located to the west of the Hanford Site in Washington state, as well as the Nez Perce Reservation in Idaho, and the Confederated Tribes of the Umatilla Indian Reservation inhabited by the Umatilla, Cayuse, and Walla Walla in Oregon. The Wanapum did not demand federal recognition, but they remained on their traditional territory in what became Hanford and embraced their role as caretakers of the land and as a People who are taken care of by the land in return.[9]

The website of the Confederated Tribes of the Umatilla Indian Reservation (CTUIR), cohabited by the Cayuse, Umatilla, and Walla Walla, articulates Indigenous Peoples' strong ties to the local geography. According to their explanation, what was created first was water, followed by life and land; as "land promised to take care of all life, all life promised to take care of the land."[10] It is interesting that Indigenous narratives of so-called creation often track with the stages of planetary development described by geologists. While the abundant fisheries were the staple of all life in the region, "eagles, bears, coyotes, cougars and Indians were amongst those who relied on the salmon."[11] Native people are listed along with other beings who rely on the salmon. The Confederated Tribes' website does not draw a clear line between humans and other species who are all seen as intertwined with all life, land, and water. Similarly, Nez Perce tribal citizen Allen Pinkham explains:

> Sometimes I try to get people to compare plant and animal species with their own body parts. For instance, the buffalo could be a finger, the passenger pigeon another finger, the peregrine falcon another finger, the

wrist could be a sockeye salmon. If you relate these body parts to these species, how many would you eliminate before you would say, "Stop"? You can get along pretty well if you lose a finger, but if you keep doing that, when is it enough? I learned this philosophy from my elders. Even Joseph himself said, "I am of the earth." Well, if you consider yourself part of the earth, you won't sacrifice those body parts.[12]

The treaties negotiated between the tribes and the US settler state reflect long-standing relational histories in and of that place in addition to the imposition of newer settler property relations. Prior to settler encroachment, human and more-than-human relations had transformed the area's landscape over generations. Within the treaties, tribes negotiated to retain what we can call ecological resource rights. The local tribes' treaty with the United States, concluded in Walla Walla Valley in 1855, for example, recognizes and protects "the exclusive right of taking fish in the streams running through and bordering said reservation" and also secures "the privilege of hunting, gathering roots and berries, and pasturing their stock on unclaimed lands in common with citizens."[13]

It should be noted that even though the relationship at the time between tribal delegates and the US government was absolutely unequal, the tribes asserted their rights to access natural and cultural resources in legalized terms within the treaty. While settler-colonial powers did extensive damage to Indigenous livelihoods by seizing much of their traditional territories, relocating Indigenous Peoples to the reservations, and subjecting them to physical and cultural genocide, the tribes—in part aided by recourse to legal protections afforded in treaty—continue to assert inherent sovereignty and forge ongoing coconstitutive social relations and structures with/in their homelands.

Tribal officials and policy makers today also speak in ways that are consistent with their ancestors' ideas as reflected in treaties. In interviews conducted by Noriko Ishiyama, tribal representatives involved in environmental policy making emphasized their rights to the land occupied by Hanford—land that historically provided their Peoples with food, medicines, and kinship relations with various animals, plants, and other more-than-human entities such as rocks and water. We could say that such human-place relations sustained the Peoples both physically, in terms of political economy, and shaped their very Peoplehoods.

Manhattan Project and Destruction of the Environment

The establishment of a military zone at the Hanford Site in the service of US national security posed significant ecological, political-economic, social, and cultural threats to the tribes. In the *Final Report: Hanford Tribal Stewardship*, prepared for the Nez Perce Tribe, the tribes' connection to the Hanford land is explained:

> The area that encompasses the Hanford reserve continued, up through 1943, to provide the tribes with traditional foods and medicines that were harvested throughout the year. Deer and elk augmented the salmon, and supplied meat, clothes, and tools. In 1943, with the establishment of Hanford, the government restricted the ability of the Tribes to exercise the rights guaranteed them under the treaties. The Tribes continue to view all of Hanford as a cultural reserve with abundant natural resources and critical habitats as well as many sites of significant historical and spiritual importance to the Yakama, Umatilla, Wanapum, and Nez Perce peoples.[14]

While the land appropriated for the Hanford nuclear reservation had been shaped and cultivated for centuries by the relations of multiple beings—species, water, and land—US military officials understood the area from within a fundamentally different conceptual framework and worldview, and without the historical knowledge of Indigenous Peoples who for generations called this place home. Military officials had selected this place as a site for the top-secret Manhattan Project because they saw it as "an isolated waste land, remote from population centers."[15] Throughout World War II and the Cold War, Hanford fulfilled a mission to support atomic weapons production for the national security of the US settler state. Eventually the Hanford Site became one of the most contaminated sites on the planet. Hanford was added to the National Priorities List in 1989, when the US Department of Energy, US Environmental Protection Agency, and Washington Department of Ecology started to work together to attempt to clean up the site.[16]

Because the Hanford Site was created for secretive military purposes, the US federal government kept information regarding ecological and health risks secret. Violent and totalizing notions of sacrifice are key to militarism. At Hanford, the government disregarded the welfare and health of local people—both Indigenous and non-Indigenous. Living downwind as well as downstream of the nuclear site, human and more-than-human inhabitants have been subjected to high levels of radioactive pollution over decades that have produced significant effects, including tons of leftover plutonium, contaminated ground water, and leaking waste tanks.[17]

In addition to the human health effects that the broader local population has been subjected to from environmental devastation, Indigenous Peoples have had their cultural practices damaged. Their everyday lives and traditions came into being in intimate relationship with the local ecological system. In the report *Facing Cancer in Indian Country: The Yakama Nation and Pacific Northwest Tribes*, Christine Walsh, project coordinator for the Contaminated Subsistence Fish Project of the Indian Health Service, explains:

> One of the things that is very difficult for outsiders to understand is the relationship between the fish and the people. The tribal people will tell you that they are the salmon. Salmon is the first food for many of the babies born here, and from birth to the death of the body, the people and the salmon share each other's lives. Many religious and cultural practices are centered on salmon.[18]

Here Indigenous Peoples approach salmon without a human-nonhuman hierarchy. This quote illustrates the dynamic co-becoming of beings who are profoundly connected throughout the development of their life courses in this particular place. The nuclear projects pursued at Hanford thus devastated both Indigenous Peoples and salmon bodies, as well as their cultural, social relations and Indigenous livelihoods.

Hanford Land and Wanapum Co-becoming

The impacted tribes have not been silent victims in the development of nuclear weapons at Hanford. Although they did not consent to the processes of plutonium production, they have been at the decision-making table for cleanup and post-cleanup use of land at the Hanford Site. Their efforts to participate in environmental remediation and policy making have aimed to recover ties to cherished lands. Indigenous Peoples' regard for and responsibilities to the landscape appropriated for the Hanford Site have not diminished over the generations of settler-colonial environmental violence. And their perceptions of Hanford history and related policy making do not necessarily coincide with the perceptions of the site among mainstream environmental protagonists.

Wanapum leader, Rex Buck, explained in an interview that Hanford is a "blessing in disguise."[19] According to Buck, Hanford is a sacred place that invited the nuclear project in order to protect the land from the further invasion and development of the settlers. As a result, the place has been contaminated, but at least on the surface it still presents a landscape before Euro-American settlement, precisely because access has been prohibited for security reasons. Indeed, the large segment of the land occupied by Hanford and its buffer zone does not have settlers' homes, restaurants, schools, or cafés. He asserts that this sacred place will heal itself, and the Wanapum will be part of this healing process as caretakers of the land. He does recognize the severe impacts of radioactive contamination and has made sure to attend numerous meetings to be part of Indigenous struggles to pressure the federal government to clean up the contamination. But he simultaneously believes in the agency of the land to recover itself. The Wanapum will always be there to recover with the land.[20]

Thinking in terms of co-becoming, Rex Buck's analysis is perhaps unsettling to those accustomed to more usual binary ethical language that tends to oppose good and bad relations, pristine and contaminated landscapes. Can one speak of "place/space, relationality and a more-than-human, material co-becoming" when deep trauma is involved from land appropriation, nuclear weapons production, and the disruption of traditional livelihoods and cultural practices from extreme contamination?[21] But certainly co-becoming sometimes involves trauma to both humans and more than humans, both physical and historical trauma. We sometimes continue to relate with troublesome human relatives, perhaps with little choice, but with hope for an improved relationship over time—if not in this lifetime, then in future lifetimes or generations. Unfortunately, not all relations are good relations, not all are evenly reciprocal. Even though the land may be damaged, now dangerous and no longer able to provide for the Wanapum, Rex Buck declares that the Wanapum will continue to care for the land. How many of us have made these kinds of difficult choices (or not quite choices) with human relations?

The Wanapum leader's story resonates with the experience of Russell Jim from the Yakama Nation. Jim is well known across Indian Country for pressuring the federal government for decades to clean up the Hanford Site. In the 1980s, Jim was attending a meeting where policymakers and scholars discussed how they would be able to reliably inform humans in the distant future, possibly 100,000 years from then, about the danger of high-level radioactive waste buried underground in our lifetime. When discussions were going nowhere, Jim told meeting participants not to worry, because "We'll tell them."[22] He meant that his people, the Yakama, would tell future generations about the radioactive waste, that it was not something precious but something extremely harmful to their health. Jim's message was clear: Indigenous Peoples have existed in the Hanford area from time immemorial, and they will still be there to fulfill their responsibilities to care for the land in 100,000 years.[23]

Such "future scenarios" conversations that took place in the late 1980s and throughout the 1990s, early on in the cleanup and management of the US nuclear weapons complex, were fantastical and yet mundane. Kim TallBear, freshly graduated with a master's degree in environmental policy, took part in such discussions held at meetings across the country but especially near to nuclear facilities in Washington state and Idaho. Meetings populated by federal agency employees, tribal representatives, and private contractors (whose companies were paid exorbitant contracting fees) took place in carpeted hotel and agency conference rooms that all looked the same. Over coffee and pastries and with plenty of mind-numbing federal terminology, both non-Indigenous and Indigenous citizens, scientists, bureaucrats, and Indigenous traditional knowledge holders speculated about how present-day humans might reliably transmit knowledge about dangerous radioactive material buried underground five thousand generations in the future.

How could this be done when multiple civilizations would have come and gone, and the languages of our time would likely be indecipherable to future societies? Yet the very long view of history expressed in these paradoxical late-twentieth-century bureaucratic scenes by Russell Jim and other Indigenous thinkers is interesting to consider. Imagine the implications for development of a much longer historical view. Imagine accountability not only to a present-day nation-state or electorate but to relations across time. The environmental and human tragedy of nuclear weapons development cannot be undone despite US Department of Energy (DOE) efforts at remediation. But are there lessons to be learned for decision-making in the multispecies accountability of Russell Jim and other Indigenous thinkers to other species/beings now and in the future? We doubt an imperial US state is capable of learning from such lessons, but relational concepts like co-becoming may help other communities and societies, those that work to regroup after a US empire in decline. We need to cultivate different (not necessarily new) narratives to guide us in social planning and policy making after settler hierarchies of life fail us, as they are doing.

Environmental History of Skull Valley

Another "national sacrifice zone" lies nearly seven hundred miles southeast from the Hanford Site in Washington state in Tooele County, Utah. Therein lies the forty-mile-long Skull Valley and within that the Skull Valley Band of Goshute Indian Reservation. Skull Valley has been thoroughly contaminated by US Cold War military enterprises.[24] Mike Davis once posed the following question: "Imagine for a moment that you are a rogue plutonium atom, a homeless anthrax spore or a vagabond dioxin molecule. Where would you most like to live?" According to Davis, "the answer, as every little toxin and bacterium knows, is Tooele County, Utah. Here, just west of Salt Lake City, is the nation's greatest concentration of hyper-hazardous and ultra-deadly materials. The Nuclear Test Site in Nevada is a nature preserve compared to the poisoned landscapes of Tooele County."[25]

Given this environmental history and present-day reality of Skull Valley, the reaction by the predominately Mormon, non-Native community to the Skull Valley Band of Goshute Indians' interest in hosting an interim nuclear waste storage facility is both fascinating and troubling. While tribes in nearby states visibly influence local geographies, Forrest Cuch, then-director of Utah's Division of Indian Affairs in 1999, described Utah as a "monolithic cultural landscape dominated by one religion," leaving little room for Indigenous cultural influence.[26] It is approximately a seventy-mile drive from the Skull Valley Band of Goshute Reservation to Temple Square, the headquarters of the Church of Jesus Christ of Latter-day Saints, which sits in the core of the urban grid landscape of Salt Lake City. White Mormon settlement has effectively erased Indigenous-land relations for more than a century. The Skull Valley Band, composed of approximately 130 tribal members, of which less than thirty actually live on reservation, was rendered invisible in settler history and in the political economy of the state of Utah.

Before Mormon settlement, the Skull Valley Band lived in an area reaching from present-day Salt Lake City to Skull Valley—which is located approximately forty miles west of the city. Although they were largely culturally overrun in large-scale, Euro-American and Mormon settlement in the 1840s, the Skull Valley Band of Goshute resisted relocation to the Utah-Nevada border. In 1917, an executive order approved federal recognition for the tribe. Predictably, the land officially designated as the Skull Valley Band of Goshute Reservation consisted of only a small parcel useless to the settler farmers and ranchers.

While Goshutes had hunted and gathered and relocated seasonally to efficiently use the limited resources available in their homeland, Euro-American settlement transformed the ecological system of the desert. The settlers introduced horses and mules that overgrazed the grasses and lessened the prevalence of seeds that the tribespeople gathered. As a result, once a self-sufficient and strong people who lived with a deep knowledge of the ecological cycles of the desert, the Skull Valley Band became materially and culturally vulnerable due to the outsiders' invasion.

Several federal military territories surround the Skull Valley Band of Goshute Reservation.[27] Open-air nerve agent tests, as well as chemical and biological weapon tests and incineration, have been conducted on these military reserves. While the local municipalities received immediate economic benefits and job opportunities, the tribe received nothing but political and ecological risk. In 1968, more than six thousand sheep were killed by a nerve gas leak at Dugway Proving Ground. Some of their bodies were buried within the territory of the reservation without the tribe's knowledge or consent.

The end of the Cold War caused fear among the Tooele County policy makers and residents that federal money would stop flowing and their economy would fall apart.[28] In the words of Chip Ward, the co-founder of the local environmental groups Families Against Incinerator Risk (FAIR) and the Healthy Environment Alliance of Utah (HEAL Utah), "the military paved the way for private polluters."[29] Accordingly, they sought the economic and employment opportunities provided by the commercial hazardous and low-level radioactive waste industry as well as a magnesium-processing facility, which was added to the federal Environmental Protection Agency's National Priorities List in 2009.

The impact of the pollution on the local population was feared to be significant. West Desert HEAL, a local environmental advocacy group, stated that "more than 33 pounds of toxic pollution per capita is emitted each year in Utah... compared to a national average of just under 6 pounds per capita per year."[30] A local community health survey indicated high rates of cancer, respiratory problems, reproductive problems, birth defects, and other severe health problems.[31] Nonetheless, since the facilities provided tremendous financial benefits, they were tolerated and even welcomed with open arms by local communities.

The Skull Valley Band of Goshute Indians and Nuclear Waste

It is within this land-use and historical context that the Skull Valley Band of Goshute Indians announced in the early 1990s that they would begin negotiating with the US Nuclear Waste Negotiator on a Monitored Retrievable Storage (MRS) facility and then, when the federal plan reached a dead end, on a leasing contract with Private Fuel Storage, a limited liability company composed of eight electric utilities, to store spent nuclear fuel. They faced hostile reactions from many different directions. Environmentalists, environmental justice advocates, other tribes and tribal organizations, and some Skull Valley Band of Goshute members vociferously objected to the tribal government's consideration of hosting an interim storage of high-level radioactive waste.[32]

For the Skull Valley Band of Goshute leaders, however, this was not a shocking or sensational economic development move. They simply made one more decision to survive in an already devastated ecological space.[33] In other words, the Goshute leadership followed a tradition of being "survivors," utilizing available resources that included even playing host to toxic molecules produced by a US settler state in their lands appropriated and dominated by both military and industrial development. According to the research conducted by Danielle Endres, the Skull Valley chairman "used the perception of the land as wasteland to justify the proposal," referring to his reservation "as 'a wasteland—a beautiful wasteland.'"[34]

In contrast, some tribal members who identified as traditionalists networked broadly with both Indigenous and non-Indigenous environmentalists and environmental justice advocates to oppose the Private Fuel Storage project.[35] The small Indigenous community in the desert was deeply divided over the issue of land use, environmental justice, and who represented so-called authentic Goshutes. Ironically, the tribal government encountered harsh objections too from the state of Utah, despite the state's own history of welcoming numerous ecologically disastrous projects. The Skull Valley Band of Goshute chairman denounced his critics' assertions:

> People need to understand that this whole area has already been deemed a waste zone by the federal government, the state of Utah and the county. That's why we're so surprised about Gov. Leavitt's opposition. Tooele Depot, a military site, stores 40% of the nation's nerve gas and other hazardous gas only 40 miles away from us. Dugway Proving Grounds, an experimental life sciences center, is only 14 miles away, and it experi-

ments with viruses like the plague and tuberculosis. Within a 40 mile radius there are three hazardous waste dumps and a low-level radioactive waste dump. From all directions, north, south, east and west, we're surrounded by the waste from Tooele County, the state of Utah, and US society. Over 30% of the tribe is children, so yes, we're very concerned about the effects of all this.[36]

In line with the prevalent political tradition of the federal and state governments, the Skull Valley Band of Goshute leaders did not consult with neighboring communities as they negotiated a plan to host a temporary storage of high-level radioactive waste. The governor of Utah, furious with the tribe's decision, declared an "over-my-dead-body" policy to prevent this project. The Utah politicians were offended that they did not have control over land use within the tribal reservation.

In 2006, the Nuclear Regulatory Commission (NRC) issued a license for the project, while the Bureau of Indian Affairs (BIA) did not approve the lease agreement. Meanwhile, a number of lawsuits were put together by the state of Utah as well as the opponents within the tribe. In 2010, the court made a decision not to support the BIA's denial of the project. While in 2012 the Private Fuel Storage project announced that they would withdraw after years of intense legal battles, no specific policy decisions have been made at the federal level. The issue of high-level radioactive waste storage remains unresolved.

Erasure of Goshute Stories

Tribal leaders' decision to pursue the Private Fuel Storage project seemed shocking to some, in terms of its potential political-ecological impacts. It might also surprise some because of the discourse of tribal self-determination that is anathema to most US Americans, who believe that colonization is a done deal and who rarely possess knowledge of tribal collective rights and jurisdictional authorities that gave the Skull Valley Band of Goshute grounds for negotiating to host the Private Fuel Storage. Whereas various legal, political, and social obstacles prevented the project from actually being implemented, the Goshute leadership was charged with "federal crimes, ranging from making fraudulent statements to committing theft and bank fraud."[37] Eventually, locals as well as the national public lost interest in this story, which had been widely covered in the press in the late 1990s and early 2000s. Margene Bullcreek, who was a resident of the reservation and a vocal opponent of the project within the tribe, passed away in 2015. The following year, Leon Bear, who served as a tribal chairman for a decade and spearheaded the effort to host a nuclear waste facility, passed as well.

The fundamental problem grounding this controversy remains unaddressed: The tribe invited a nuclear waste facility because of severe poverty and very few job opportunities in an already ecologically compromised landscape. Indigenous experience and needs continued to be erased and unaddressed by the local, county, and state governments. The Utah state government, which has occupied, colonized, and contributed to ecological devastation of traditional Goshute lands, adamantly attacked the tribe's efforts to host the facility and yet continues to do nothing to assist the Goshute so that they would not have to consider pursuing another project, like the failed Private Fuel Storage initiative, in the future.

Forrest Cuch, then-director of the Utah Division of Indian Affairs, was put in a difficult situation in his role as the state government liaison to the Utah Indian tribes during the conflict over the Private Fuel Storage project. He nonetheless developed respectful relationships with both the tribal leadership and Goshute tribal member opponents to the project. Recognizing the complicated historical and social contexts, he maintained his support for tribal sovereignty and advocated that the tribe's decision be respected.

Recalling his challenging experience as a state employee, Cuch summed up the fundamental dilemma—that "Indians are still invisible in Utah" and that for the majority of Utahns, the Indigenous Peoples' presence and their narratives of both history and the present represented "the inconvenient truth" that Utahns were unprepared to deal with.[38] The "inconvenient truth" of Indigenous Peoples' experiences in Utah is a discomforting challenge to whites with their romanticized historical narratives of hardworking settlers who cultivated empty and untouched lands.[39] Such violent eradication of Indigenous presence, along with countervailing Indigenous historical interpretations, spurred the Goshute leaders' decision to pursue the Private Fuel Storage project for survival. For those leaders, to survive as a People meant that they must find ways to retain their collectivity together in place and thus their culture in a society dominated by settler colonialism.

Not All Relations Are Good Relations

We end by continuing our familial metaphor of dysfunctional relating and the project of recuperating relationships, despite uneven relations of care, to analyze the standpoints of two Indigenous communities and two specific radioactive waste cases. Recall Wanapum leader Rex Buck's morally complicated relationship with the DOE and the Hanford Site. Despite extensive radioactive contamination that will animate the land for tens of thousands of years, Buck called the place sacred. Obviously, his idea of sacred is not synonymous with purity. He looked to a future when the land would heal. He asserted Wanapum caretaking of that space for a timeframe that makes seven generations look like seven months. Buck clearly learned from experience with his more powerful and sometimes-violent relation, the US Department of Energy, to work within dysfunctional kinship and literal toxicity as he tries to stay a committed relative to his home territory. Both the land and the Wanapum have been vulnerable to the DOE. Buck recognized the physical and psychic impacts of radioactive assault, and he attended various meetings where the Wanapum and other Indigenous Peoples encouraged federal agencies to get clean after their long assault on the Hanford Site and the planet. Buck's participation in this difficult relationship is an act of faith that the land, and with it the Wanapum, can heal for future generations.

The Skull Valley Band of Goshute, with their twenty-first-century land base restricted to a toxic fraction of their traditional territory, are not living at a safe physical distance from toxicity. The Goshute, therefore, engaged in their own act of faith with yet another more powerful and in part toxic relation—not a settler government, but a corporation, Private Fuel Storage—when they invited controlled contamination in the form of a heavily regulated and potentially well-compensated spent nuclear fuel facility. With their close proximity to and immediate risk from an already toxic landscape, the Skull Valley Band of Goshute were differently inclined in their orientation to their toxic land relations than were the Wanapum, Yakama, and other tribes with cultural and historical stakes in the Hanford Site.

The Goshute were in survival mode, yet they were still being held to some romantic expectation of unconditional, Indigenous caretaking of land while they suffered material deprivation and toxic harm from powerful white governments, both the state of Utah and the US, that continue to abuse and gluttonously nourish themselves at the expense of the Goshute, the land, and more-than-human relations. In making the unromantic choice to pursue a partnership with a company that would add a nuclear waste facility to their toxic landscape, but financially compensate them, the Goshute pursued one of the only choices they had to help them withstand ongoing abuse from the state of Utah and the US federal government. The Skull Valley Band of Goshute simply hoped they might survive for a future in which their descendants had better choices.

Notes

1. Coffey and Tsosie, "Rethinking the Tribal Sovereignty Doctrine."
2. Whyte, "Settler Colonialism," 126.
3. Davis and Todd, "On the Importance of a Date," 769.
4. Bawaka Country et al., "Co-becoming Bawaka," 455.
5. Shorter, "Spirituality," 433.
6. Shorter, "Spirituality," 433.
7. Coffey and Tsosie, "Rethinking the Tribal Sovereignty Doctrine," 196.
8. TallBear, "Identity Is a Poor Substitute for Relating," 474, 476.
9. Rex Buck, interview with Noriko Ishiyama, Richland, Washington, August 3, 2015; Rex Buck, interview with Noriko Ishiyama, Mattawa, Washington, September 7, 2017.
10. Confederated Tribes of the Umatilla Indian Reservation, "A Brief History of CTUIR."
11. Confederated Tribes of the Umatilla Indian Reservation, "A Brief History of CTUIR."
12. Landeen and Pinkham, *Salmon and His People*, ix.
13. Governor's Office of Indian Affairs, "Treaty of Walla Walla, 1855."
14. Baptiste, *Final Report*, 3–4.
15. Harvey, *History of the Hanford Site*, 4.
16. According to the US EPA (Environmental Protection Agency) website, the National Priorities List is "the list of sites of national priority among the known releases or threatened releases of hazardous substances, pollutants, or contaminants throughout the United States and its territories." United States Environmental Protection Agency, "Superfund: National Priorities List."
17. Columbia Riverkeeper, *Hanford and the River*; Gerber, *On the Home Front*; Grossman, Nussbaum, and Nussbaum, "Thyrotoxicosis among Hanford, Washington"; White, *Organic Machine*.
18. Reuben, *Facing Cancer*, 1.
19. Rex Buck, interview with Noriko Ishiyama, Richland, Washington, August 3, 2015.
20. Ishiyama, *"Giseikuiki" no America*, 46–48.
21. Bawaka Country et al., "Co-becoming Bawaka," 469.
22. Russell Jim, interview with Noriko Ishiyama, Union Gap, Washington, August 11, 2015.
23. Ishiyama, *"Giseikuiki" no America*, 222–23.
24. Kuletz, *Tainted Desert*; Endres, "Sacred Land or National Sacrifice Zone"; Ishiyama, *"Giseikuiki" no America*; Shumway and Jackson, "Place Making, Hazardous Waste."
25. Davis, "Utah's Toxic Heaven," 35.
26. Ishiyama, "Environmental Justice and American Indian Sovereignty," 124.
27. Ishiyama, "Environmental Justice and American Indian Sovereignty."
28. Shumway and Jackson, "Place Making, Hazardous Waste," 444–45.
29. Ward, *Canaries on the Rim*, 47.

30 United States Nuclear Regulatory Commission, "Scoping Meeting for Preparation of an EIS for the Private Fuel Storage Facility," 43.
31 Ward, *Grantsville Community's Health*.
32 Clarke, "Construction of Goshute Political Identity"; Ishiyama, "Environmental Justice and American Indian Sovereignty"; LaDuke, *All Our Relations*, 104–6.
33 Bear, interview with Noriko Ishiyama, Salt Lake City, Utah, June 14, 2000.
34 Endres, "From Wasteland to Waste Site," 929.
35 Clarke, "The Construction of Goshute Political Identity"; Ishiyama, "Environmental Justice and American Indian Sovereignty."
36 Qtd. in Hanson, "Nuclear Agreement Continues U.S. Policy of Dumping on Goshutes."
37 Smeath, "4 Goshutes Charged with Fraud."
38 Forrest Cuch, interview with Noriko Ishiyama, Roosevelt, Utah, August 19, 2017.
39 Ishiyama, "*Giseikuiki" no America*, 207–8.

10

Multispecies Mediations in a Post-Extractive Zone

Zsuzsanna Ihar

10.1 (*previous page*) Original drawing by Feifei Zhou.

THE KHAZRI, OR NORTHERN WIND, was tearing through downtown Baku, Azerbaijan, when I first encountered him. A harbinger of dust, heat, and scattered particles, the gust made it almost impossible to spot the outline of the suspicious canine. By the time I could properly register him, albeit with eyes full of debris, his back was already arched, jaw clenched, and the glint of yellowed fangs visible. He barked with an assertiveness that, in my mind, denoted annoyance rather than aggression. It rang out across the idle-brownfield land bridging the newly constructed stone-clad buildings of the White City (Ağ Şəhər) and what little remained of the Black City (Qara Şəhər) with its oil-contaminated soil. Perhaps it was a warning against trespass, or a firm statement of territoriality, ignored. I continued along my path, less concerned about the lurking canine than dodging the windswept shrapnel billowing around me—until I felt the sting of teeth sinking into my flesh. The bite signaled that I had crossed a line and failed to notice important signals. A rush of blood and adrenaline mediated my response as I quickly tried to assess the situation. I had been stubbornly forging ahead, forgetful that acts of respect sometimes look and feel very similar to wariness and hesitation. In an Anthropocene condition, where everything seems to be tinged with urgency and the need for a quick response, respect may reveal itself a counterintuitive, cautionary practice.

My canine encounter gave way to the usual reliance on the state for the provision of emergency care: hospital visits to tend to the wound, vaccine shots for rabies and tetanus, travel insurance claims, and the documentation of injury via shaky iPhone shots and clinician reports. Weeks later, bandaged up and slowly healing, I thought to myself: if a stray dog exercises such will and compulsion to protect a seemingly barren wasteland—of toxic soil and industrial debris—he must recognize something of importance and value that is unintelligible to me. It must be a place he frequents with intention. As Alfred North Whitehead observes, "The primary glimmering of consciousness reveals something that matters."[1] On the level of common sense, the dog bit me because I had strayed into its territory—a space vulnerable to intrusions from competing stray packs, real-estate proprietors, animal control agents, and other unfriendly (even antagonistic) subjects.

Passing by the industrial zone a few weeks later, I spotted a pregnant partner alongside the now familiar (yet still threatening) dog, suggesting to me that the attack was partly inspired by its responsibility and duty to others. Perhaps my exclusion strengthened social relationships and kinship, with the bite demonstrating the dog's capacity to maintain the integrity of a decidedly nonhuman, even antihuman, social structure. It was a structure which required no companionship nor approval from humans. On a more symbolic level, the bite was also a strategic move by the stray to resist what I represented, and perhaps heralded, with my very presence in the Black City. I was part of a cosmopolitan project of deindustrialization and urbanization. A global aesthetic was being cultivated here as a sovereign nation-state reimagined itself through projects of environmental remediation. As a human, a foreigner, a tourist, and a scholar, I too was complicit in multispecies mediations in Baku's former industrial zones (see figure 10.2).

My encounter with the stray dog offers an opportunity to reflect on alertness and attentiveness. I understand the bite as a moment of multispecies mediation rather than establishing a jurisdiction or exercising judgment.[2] What counts as the right or wrong action—and on whose part—is unclear. On the one hand, the bite might be read as an undeserving act of violence committed against an innocent passerby. But it might also be understood as a recalibration of necessary boundaries and distance—an unjust act that reinstates some kind of territorial and relational justice, for some and against others. The encounter opens speculative spaces for rethinking whom justice is for, and who gets to deliver it.

Dominant models of justice tend to limit the range of action to acts of intervention, reparation, compensation, or restitution by institutions.[3] Justice, in this sense, emerges from powers that "claim and hold the rights to articulate it," in the words of Zygmunt Bauman.[4] This vision reduces complex webs of practices, relations, and modes of conduct to simplistic judgments of right or wrong, aspiring only to balance the checks and reintroduce a sense of governability and assumed harmony.[5] Judicial institutions often aim to

10.2 Culprit behind the Ağşəhər incident, spotted a week later in Baku White City, July 18, 2019. Photograph by Zsuszanna Ihar.

restore an original or ideal state—aspiring only to meet demands, smooth over disagreements, get rid of irritants, and pacify sources of complaint or resistance.

Dogs who live with people in Western urban environments often live within clear dominance hierarchies, where judgments of right and wrong by humans can seem oppressive or putative. Donna Haraway's reflections on positive bondage in human-canine training practices illustrates how limitations on freedom allow for a mutually beneficial sharing of space. It is worth noting that Haraway's interactions with her dogs have been structured by municipal and state jurisdictions, where such restraint and training is mandated. Her *Companion Species Manifesto*[6] is also informed by practices in management sciences, that place a premium on mastery, command, and control. The human sets up the conditions of these canine encounters. The former industrial zones of Baku, on the other hand, remain ungoverned, allowing for a wider range of improvisation, negotiation, and mediation. Within the context of my painful encounter, I was not the one setting the rules. The dog and I were encountering each other in a space neglected by the sovereign state, where human mastery or even safety was not guaranteed. It was a spent and expired space where new protocols could slowly emerge.

Spaces that fall outside the jurisdiction of governments, or that are beyond the purview of global capitalist enterprises, can give rise to interspecies relations that breakaway from anthropocentric tenets about family, citizenship, and appropriate affiliations. The undomesticated and unmastered dog, as Carla Freccero notes, occupies the "periphery of an observer's known world,"[7] embodies a hostility which requires a more careful and tentative approach. It resists the normative world of humans, where predictable social cues produce unproblematized connections. Similarly, Harlan Weaver's work on "intimacy-without-relatedness"[8] shows how human-canine relationships need not be reciprocal nor characterized by positive qualities of love and attachment in order to be valued or considered meaningful. For Weaver, the marginal space of the shelter and the socially reviled figure of the pit bull conjure up a different multispecies world, in which inhumanity becomes foundational for coexistence and shared political duty. Dogs who are assumed to be part of criminal worlds demand different articulations of justice.

Conflicts in multispecies worlds might be reframed along the lines of Carol Gilligan's morality of care, where justice and injustice constitute issues of relation and response, in turn reliant on the maintenance of specific forms of sociality, as well as open-endedness.[9] Gilligan conceives relational and care-oriented forms of justice as incommensurable with contractual and hierarchical frameworks. When it comes to multispecies issues, however, we should not be so quick to do away with forms of relation that maintain categorical differences, boundary lines, and pecking orders, as well as unresolvable conflicts. It would be a mistake to construct moral or ethical problems as conflicts between an individuated self and a singular other, to be resolved by the eradication of difference and the attainment of a harmonious outcome.[10]

In post-extractive, postindustrial spaces like Baku, the very notion of justice brings into the picture all kinds of species boundaries and exclusions—those that are valued and those that are not. Attending to the ways in which historical, cultural, and material contingencies arrange life offers possibilities for forging alternate alliances. In these zones, the past violence of extraction has left a lineage of disrupted processes—muddying efforts at establishing appropriate affiliations and social organizations. Here animals are making their own articulations of "division and alterity,"[11] to borrow a phrase from Joanna Latimer. Shifting the focus away from the pursuit of rights or representation for individual animals or plants offers an opportunity to welcome spontaneous interruptions (even if they are violent or uncomfortable).[12] Haraway and many others have long insisted that multispecies worlds are not places of Edenic peace.[13] The damages wrought by human systems of capital, power, and property cannot be healed by the resurgence of nature, or the easy reintegration of diverse flora and fauna.[14] Instead, it is important to take account of and reckon with distributed possibilities for justice in ecological assemblages.

The multispecies mediations I examine in this chapter offer a window into worlds at war.[15] Remediation and ecological restoration projects launched by the Azerbaijani state aim to correct the social and material injustices caused by decades of continuous oil production and processing. With the partial conversion of Baku's Black City into the much-lauded environmental district of the White City, select plants and animals are being recognized as prospective pacifying agents. Hopes are pinned on these introduced organisms for resolving environmental discord and smoothing out boundaries and divisions between humans and other species—in effect, depoliticizing relations. These nature-making interventions frame environmental wrongdoing as moral debt and herald restitution through idioms of similarity, beauty, and compatibility (sovmestimost').[16] The state is striving to exert a legitimate monopoly on environmental justice—one that instrumentalizes particular multispecies worlds to further the management and eventual displacement of human and multispecies populations deemed undesirable, ungovernable, and uncharismatic.

As I studied the dreamworlds of the White City, I also dwelled in the Black City where there are makeshift houses, weedy garden beds, and industrial relics from thirty years ago. Within this environment, I listened to the stories of internally displaced people (Məcburi köçkün) from the provinces of Ağdam, Füzuli, and Lachin. They talked about their own encounters with stray dogs and noxious weeds—specifically the camelthorn (Alhagi maurorum).[17] In the Black City settlement, I found important breaks, pauses, and hiatuses in human and multispecies relations.[18] The "broken earth" (parçalanmış torpaq) of the Black City acts as a fault line where exclusion and conflict play out in communal politics between species. The presence of meaningful difference when it comes to habits, behaviors, and ideas of a good life unsettles the default conceptualization of relation as one of positive mutuality and connection. In the Black City, I found epistemic and ontological differences between species that were the very linchpin of everyday care, conduct, and modes of sociality—an alter-relation (or alteration) rather than an anti-relation.

In the Black City, there are practices of justice that exist without the necessity of peace, produced by what Anna Tsing calls "the work of many organisms, negotiating across differences ... in the midst of disturbance."[19] This is not the realm of state jurisdiction with its clear-cut resolutions or its focus on representing the interests of self-contained beings. Rather, the blasted landscape of the extractive industry is one that is already composed of a bundle of arrangements, dynamics, relations, and different scales of mattering.[20] It is an environment best conceptualized through Bettina Stoetzer's ruderal perspective, which takes rubble and ruin not as markers of universal hybridity, but as evidence of minor disruptions within the urban fabric.[21] Stoetzer's scale is one of small gardens, small gaps, and small interruptions, which, despite their scale, matter at the level of nation, race, and capital.

Riffing off Stoetzer, I found that Baku's peripheral spaces were sites of small justices, reliant on incremental shifts and slight alterations in the urban fabric, rather than wholesale transformations. These were spaces created through *multispecies mediation* rather than *environmental remediation*. Returning to the etymological antecedent of *mediare*, to "be in the middle,"[22] I understand mediation as an opportunity to shift away from measured responses and toward necessary disputes.[23] By staying in the middle—between worlds—I found that mediations produced cosmopolitical possibilities, to build on the ideas of Isabelle Stengers. The middle maintains difference and allows for an intermediary space to emerge. In this space, tactics are divisive, methods cautious, and connections not always successfully forged, or indeed desired. Instead of discovering enduring entanglements, I unearthed a stray justice that bares its fangs, at once inviting recognition and demanding distance.

The Black City's Oasis

One of my frequent rituals in Baku was to ride the bus down Nobel Avenue. I would start my journey at the headquarters of SOCAL (The State Oil Company of Azerbaijan Republic) on the Bayil side of the boulevard and get off in the White City. The road curved along the Caspian Sea which, by late afternoon, was all oil slick and pinkish haze. I'd find myself walking past a large marble monument, erected in 2006 to commemorate the "Comprehensive Action Plan for Improving Ecological Conditions in the Republic of Azerbaijan during 2006–10." Alongside stating the key promises of the White City's extensive 221-hectare remediation, the monument featured a quote by late-president Heydar Aliyev: "The Black City throughout centuries will turn white, clean, there will be grown flowers, and it will come to be a beautiful sight of Azerbaijan." This quote exemplifies the repurposing of the environment as a potent tool of state-building. In this nature-making project disguised as repair, the scientific practices involved in landscaping and horticulture have simultaneously become prospective tools of mastery and atonement over an unruly and often ungovernable post-extractive space—an environment repurposed for state-building.[24] In order to realize a new cosmopolitan Baku, amid a legacy of heavy industrialism and ecological exploitation, the state needed to import commodities, people, and a host of plants and animals they perceived as appropriate.

Over the months I had spent visiting the White City, ornamental perennial flowers bloomed at a rate which seemed to match the incremental disappearance of the miscellaneous allotments of invasive weeds. Well-groomed and purebred dogs began outnumbering the strays which had once accompanied hooded crows in the daily scavenge for leftover food and the occasional rodent. The middle-class residents of flagship eco-friendly buildings moved through the former wasteland with vitality—starting community gardens and compost heaps, decorating their balconies with an assortment of edible plants, and exchanging tips regarding permaculture practice. The White City in many ways sought to redeem a country that had been routinely portrayed as one of the worst offenders of petroleum-affiliated environmental destruction, with toxic soils and feral occupants haunting the national imaginary like incriminating evidence.[25] Here, justice was to be realized through the literal breaking down of matter.

This vision of justice was perhaps most apparent in the rhetoric of local environmental scientists. One of them was Ruslan, a professor of geography and environment responsible for a number of remediation action plans in Baku.[26] At our first meeting, Ruslan led me to a noticeboard flagging a number of "uncooperative" (neposlushnyy) species. Alongside genus names written in thick, red ink, there was a comprehensive map marking out suspected habitats and visual evidence of unruly and anti-social behavior—for instance, upturned nests, wounded livestock, scattered rubbish, destroyed garden beds, foaming mouths, damaged property, and epiphytic growths. In contrast to the prevalent conceptualization of ferality as a proliferation beyond human management and control, Ruslan registered ferality as something that prevented relations in multispecies worlds. He stressed the importance of rekindling relationships with creatures decimated and driven to near extinction by the industry but equally the need to introduce a new cast of benevolent characters. Ruslan described this process of species introduction as "matchmaking" intended to bring "compatible" (sovmestimyy) flora and fauna together.[27] Beings exemplifying values of peacefulness, cooperation, and nonviolence were spoken of lovingly, with Ruslan arguing that "nature when harmonious should make you forget whether you are human or not... there needs to be a pleasant co-existence, an overcoming of age-old division." Ruslan's ideal species shared a number of characteristics. For instance, they tended to respond well to external maintenance. They did not harbor the type of disease or vermin capable of jeopardizing the residents of sites under development. And they were (more often than not) aesthetically pleasing.

A foundational element in achieving the post-extractive justice described above has been the framing of nature as a scientific enterprise—one that necessitates expert opinion, institutional involvement, economic investment, and specialized knowledge. According to these tenets, the return of nature is a project that involves reorganizing the living world. In Baku, environmental scientists and urban planners are dismantling and neutralizing previously divided spheres of life. They are trying to foster a reconciliation between nature and industry, the human and nonhuman, and the archaic and prospective. An entrepreneurial class, with a newfound appreciation for an uncomplicated and abiding nature, is both consuming and producing these dreams.

The rows of mature olive trees lining the thoroughfares of the White City serve as striking examples of prematurely hopeful multispecies celebration. Reminiscent of the traditional groves seen in ancient Absheron villages like Nardaran, the olive has spread across Baku via large-scale tree-planting events, organized by Vice President Mehriban Aliyeva and other state representatives.[28] In one instance, 650,000 trees were planted during the span of a single day to mark the 650th anniversary of the birth of Azerbaijani poet Imadeddin Nasimi. In this and other instances, vegetal beings were seen to revive good relations, collapse difference, and act as conduits for lost pastoral ties as well as poetic inspiration. However, both the rhetoric of abundant growth and the return of an "original plant made for Absheron soil" (in Ruslan's words) enabled the obfuscation of the many systems of management and intervention required for the survival of the olive trees. Despite local varieties costing half the price, most of the trees planted in the White City were purchased from high-tech agricultural enterprises using transnational seedlings.[29] One of these enterprises is Baku Agropark, a resource-intensive industrial estate that appropriated hybrid nature-making strategies from transnational plant nurseries. The rest of the seedlings were sourced from overseas industrial nurseries. Specialists were needed to ensure proper adaption to the local climate.[30] The olive tree now joins the cypresses, palms, bay-trees, poplars, hydrangeas, as well as white lilacs of Baku—removed from their lifeworlds and incorporated into systems of knowledge and capital production, all the while cloaked under the promissory image of ecological flourishing and mending.

During my fieldwork, Baku natives, or Bakinets, described the new multispecies communities (composed of charismatic species like the olive tree, rose-ringed parakeet, and elder pine) as a return to a bygone "friendship" (dostluq) between humans and other species.[31] These multispecies communities conjured up memories of a Soviet Baku which preceded the decline of industrialism in the Black City and the emergence of urban wastelands. It was a time defined by a cosmopolitan (Russophone) culture directly reflected in the manicured arrangement of the natural landscape and in ideals of a functioning and fair society. This period also predated the alleged loss of the city's unique urban identity as a result of the arrival of "rural migrants" deemed to be, in the words of one interviewee, as "feral as the stray dogs and cats that roam around Baku."

Environmental remediation in the White City has resulted in the demolition of numerous makeshift settlements in the neighborhood, inhabited by internally displaced people. This monolithic venture has accelerated what is essentially the cleansing of valuable land, near downtown and the seaside boulevard. The resettlement of the chushki (Russian slang for an internal refugee) has produced a naturescape for the upper and middle class, commercial businesses, and Western tourists. Another interviewee, who had recently moved into a newly constructed apartment building in the White City, observed that "with each weed removed, we see less poverty and waste… it makes for a more equal community, where we can all talk to one another and come together." His take essentially linked remediation to a project of justice for Bakuvians that operates through the breaking down of relations with and for those inhabiting the margins.

The systemic removal of resistant populations today harks back to the very first instance of greening in the Black/White City—the creation of the parkland surrounding Villa Petrolea and its residential suburb by the Nobel Brothers Petroleum Company.[32] Having amassed great wealth from the production of artillery missiles, cannons, hydraulic presses, and gun carriages, Ludvig Nobel (along with his brothers Robert and Alfred) established a company in 1879 named Naftaproduktionsaktiebolaget Bröderna Nobel, or Branobel for short. Branobel would come to outmaneuver American John D. Rockefeller's Standard Oil in order to lay claim to most of Baku's oilfields. With Branobel reliance on Swedish and Finnish engineers, there was a need to modify the "uncomfortable Baku" and make it ecologically (as well as climatically) palatable for the Scandinavian workers.[33] This was to be realized through the creation of lush and sprawling greenery as part of the residential suburb. Already occupied by Keşlə's landowning peasants, the site of Baku's "green miracle" encroached on the boundaries of land essential for the everyday livelihood of peasants.[34] After the eventual signing of a lease, the park remained a tightly restricted oasis, only to be used by foreign and senior employees.

The Nobel brothers—among the world's richest men during the late nineteenth century—were instrumental in displacing human communities to create parklands and in extensively reengineering the entire ecosystem.[35] Large quantities of freshwater were transported in oil tankers from the Volga River to irrigate extensively the naturally arid landscape. Later, condensed steam, generated in refineries and routed to the park via special pipelines, came to replace river water. The industrial area became a seeming oasis which could be turned into evidentiary proof of an environment repaired: the "woodless and almost fruitless piece" of the peninsula becoming a better multispecies world.[36] In many ways, the transformation of the Black City into the White City is an Anthropocene fable, where the ghosts of empire and its many frontiers return under the utopic guise of a curative nature and a relational aesthetic. It is a story where lawless and suspicious subjects—stray dogs, weeds, oil particulates, or internally displaced people—are disciplined not by the human world but by its multispecies accomplices. Life deemed nonrelational and antisocial disappears amid the greenery. Here, the dirty work of an extractive and capitalist nature is redeemed by an array of flora and fauna that promises to deliver justice in, and through, abundance.

Amid Broken Earth

My bandaged foot garnered sympathy from Ruslan and the Bakinets of the White City, who read in it evidence of an ecology in disarray. However, my canine encounter ended up becoming the source of a running joke for the internally displaced people of the Black City. I was soon identified as the scholar who presumed to be a "dog whisperer," indulging in fantasies of cross-species communication and the dissolution of territorial lines. My attitude, it seems, was premised on the normative values of coherence, resolution, and prescriptive action which continue to define Western environmental-justice rhetoric.

In 1985, Murray Bookchin identified a juncture in the development of environmental thought. Either it would "follow the path of adaptation to the existing society or the path of revolutionary opposition."[37] While radical environmental approaches do exist, the notion of justice has most often found itself articulated through a neoliberal register.[38] Here, doing right by nature is equated to a return to an originary state of equilibrium, conviviality, and abundance. This state can be traced back to Romantic notions of unspoiled wilderness and the rightful place of humans within such arrangements, concealing within it any relationship which refuses to abide by that ideal.

Amir, one of my interlocutors, was the first person to highlight a contrasting meaning to "doing justice," in and to the industrial zone. He remarked: "You came here [from Australia] with rose-colored glasses ... these dogs don't need to be fed and patted to survive ... *we* need their caretaking, more than they need ours." While I initially thought of the dog bite as a gesture of antagonism and refusal, Amir saw the dog bite as a jolt into a position of attentiveness—one that rendered explicit the sociomaterial boundaries and protocols fundamental to the organization of life for various and variably marginal communities in the Black City.

Many of the displaced individuals I interviewed complained about the comparisons made between stray dogs and their community—comparisons that, by collapsing meaningful difference, became a key tenet in justifying their community's resettlement. As Ferman, a first-generation internally displaced man from Ağdam district, put it, "The same day they started going around the settlement, setting up traps for the stray dogs, they also started appraising our houses. Just like the dogs, they think we ruin the cleanliness of the space. That we live like animals not like city people. Next, they will start saying that we bite and need to be put in shelters." Ferman explained to me that akin to the management and culling of the stray-dogs, councilors often used concerns around sanitation and illness to warrant the demolition of houses and the movement of families to the outskirts of Baku.

As the state lumps resistant populations together, displaced individuals were pragmatically insisting on separateness, on difference, in order to fight for their right to remain within the inner city. This right has been especially tenuous since, according to the country's obligatory permit system, displaced individuals are unable to register their residence in the capital Baku and are thus prevented from owning property, legally renting, or entering any form of credit relations. Referred to as the propiska, this vestige of centralized population management was created in order to restrict the movement of individuals from rural districts into the cities. Rural migrants were effectively banned from ever attaining permanent domicile in zones deemed to be urban. This is still a powerful way to keep track of the displaced as a governable minority population, in addition to ensuring their eventual return to their native lands of Nagorno-Karabakh.

Amir, fifty-nine and technically retired, moved to the Black City in the early nineties, when his hometown of Lachin (Laçın) was occupied by Armenian forces at the height of the Nagorno-Karabakh war. With legal restrictions set out by the propiska and bureaucratic red tape making living anywhere else in downtown Baku (legally) near impossible, Amir and his family built their makeshift home on an empty allotment next to a decommissioned processing facility. Here, they joined a cast of other "troublesome" beings—including a twenty-strong pack of stray dogs. While hesitant to talk of transspecies solidarity, Amir did hint at a series of workable and livable social arrangements between humans and animals, stemming from the experience of being cast out of the so-called legitimate spaces of the city. For instance, members of the community often shared humorous anecdotes of the dogs nipping the ankles of real-estate agents or incidentally herding mischievous kids from the settlement into neat clusters. One interviewee told me that the dogs "go from being a nuisance one day, stealing chickens off windowsills... to being babysitters when the children come home from school."

Multispecies mediations involved acts of care but also more ambivalent arrangements. Residents knew to turn a blind eye when the dogs snuck into overflowing dumpsters or wandered into the open sewage channels of the settlement, which were watering holes for both the strays and kids when the temperature became unbearably hot. It is worth noting that the moments described here never sought to reform the dogs nor to bring them into a sphere of good relations with obedience and the neutralization of territorial instinct. Rather, these mediations involved embracing and making do with systems in decline, prioritizing pragmatic approaches over utopic discourses of repair and rehabilitation. They offer a path to survival in the face of violence both human and canine.

Post-extractive multispecies justice might be understood with the notion of "hospicing," proposed by Vanessa Andreotti and her colleagues. In contrast to promissory transformations, hospicing approaches unsustainable and injurious configurations (like capitalist extraction) with a palliative attitude. It sits with these systems and utilizes them as learning opportunities, allowing for different and unexpected forms of relational and world-making. Hospicing reorients the concept of justice so that it can deal with "tantrums, incontinence, anger, and hopelessness," instead of merely aiming for curative outcomes.[39] In multispecies worlds, hospicing may include instances of slow relational development, and even of outright refusal, where beings in relation won't budge beyond the category of tentative strangers. In many ways, the tentative relations between people and other species in the Black City arise due to the awareness that a greater injustice is at play.

Internally displaced refugees of the Black City live with a permanent severance between land and people, that renders dreams of sovereignty and belonging as not only fantasy but also impossibility. For Amir, the return to a "pristine nature ... in the regions" and the restoration of full citizenship is no longer an attainable outcome. It lies outside of the scope of justice. Indeed, most members of displaced communities to whom I spoke to during my time in Baku accepted that decades of war and militarization had inevitably transformed their homeland, fracturing the relationships they had forged with humans, plants, animals, and soil; however, this does not mean a permanent state of mourning, nor of exploitative depictions of violence and damage. Already a beyond-reform space, the "broken earth" (parçalanmış torpaq) of the extractive zone offers an alternative to narratives of restoration. It creates conditions where relations have to be radically reformulated to address ongoing displacement and the trauma of being cast as disposable. Here, legal attachment to land is not a prerequisite to living a good and decent life—its lack does not prevent the formation of valued relations. Indeed, for Amir, there is a pleasant mirroring between his community and the dogs. Both are reluctant to form attachments so that "goodbyes won't be too difficult" when it comes to inevitable relocation.

Feral species in the Black City ultimately multiply niches, with novelty and experimentation taking the place of laws. Interlocutors were delighted when dogs behaved in surprising and, at first, unintelligible ways. Neighbors of Amir told me of strays outright refusing to take food when offered by human hands, interpreted at first as a gesture of ungratefulness and stupidity. Instead of accepting the altruism of the humans, the strays would observe the closing time of the local butcher and wait for black rubbish bags filled with offcuts to be thrown out on the street. Unlike studies in human-dog relationships which center notions of humanness as the foundation for the attainment of rights, in Baku it was the strays dictating the terms of encounters.[40] The dogs didn't need to be "honorary humans." Instead, they made their own *mediations* while living amid human—households, enclaves, cultures, communities, social networks, and architectures to craft cunning canine ways of surviving and thriving.[41]

Internally displaced people and stray dogs both demonstrate an outright refusal of domestication. In the Black City, the stray dogs are yoldaş—an Azerbaijani word for friend or comrade, stemming from yol (road) and daş (sharer). These critters are travelling companions rather than family members, sharing a commitment to land that had been designated as uninhabitable and in need of repair. This term of address signals an ambivalent political alliance, predicated on everyday pragmatism. Justice here involves simply getting by, sharing a marginal space which has managed to evade the state and its mandate of order, hygiene, and appropriate affiliation. Amir's interaction with the strays around him is a notable example of a genuine openness to being altered and affected—to having one's own patterns of behavior challenged. Indeed, Amir advised me to notice and mimic beings and things that were the most resistant and stubborn. Through observance and emulation, I learned to make my own multispecies mediations in an otherwise difficult environment—where a protocol of difference between species ultimately ensures our mutual survival and well-being.[42]

Thinking of industrial and brownfield ecologies through the lens of alter-relation instead of anti-relation provides a potent counterpoint to scholarly narratives which depict Anthropocene environments as facilitating only processes of ruination, death, and rampant violence—a "global dump," to borrow Michael Marder's terms.[43] Through shifting the anti- to alter-, there is an opportunity to recognize the strategies of liveliness embedded in the seemingly expired sites of former capitalist activity, where bonds are forged even if they are characterized by thorniness and necessary exclusions. Instead of turning away from the ontologically toxic, the residents of the Black City show a commitment to indexing and incorporating matter which may bear the inscriptions of chemical, colonial, military, and racial violence—altering its composition until the volatility is tolerable.

Alter-relations were tangible when Amir offered me a glass of milk while also leaving a small bowl for the stray cats—to "balance out the oily molecules," in his words—or when the movements of the stray dog pack showed which zones of the neighborhood were uninhabitable. Their sense of smell indicated where the soil was too toxic to be settled and built upon. Flourishing was still possible amid traces of heavy metals as people became attuned to alter-relations with animals, plants, and chemical species. Alter-relations echo Michelle Murphy's notion of the alterlife, which indexes "the ubiquitous condition of chemically altered living-being, a condition that is shared, but unevenly so, and which divides us as much as binds us, a condition that enacts and extends colonialism and racism into the intergenerational future."[44] Within the Black City, alter-relations and alterlife are part of a post-extractive world-making which not only counters fatalistic readings of the extractive zone as a site of alienation but also refuses the utopic promises of the sovereign nation-state.

Lessons in Survival

One afternoon, while resting in his front yard, Amir stood up with his cousin and spontaneously decided to haul a full-grown camelthorn shrub out of the ground. This plant has a root system capable of absorbing vast amounts of cadmium and is a figure of alterlife, alter-relations, and multispecies mediations. The men struggled to remove the obstinate plant which latched on with its three-meter root system, known to regularly break through asphalt to the dismay of city officials. Alongside the strain of untethering the plant from the soil, the camelthorn's notorious spikes tore into the fingers of the men, causing visible scratches and even a few drops of blood by the time they finally finished with the task. Amir lay the plant on top of the plastic outdoor table, brushing off the stray dirt and debris. He pointed to its roots and said, "Look at how deep they go! How stubborn and difficult! This plant cannot be removed by a single man, nor two at times ... it just doesn't want to cooperate. It saps all the water from the ground and competes with any other plant that it comes into contact with, declaring this land to be its own. Yet, at the end of it ... we owe this plant for the survival of the community ... it has been a strange ally but one we cannot do without."

Amir's reflection was echoed by his wife, Ferize, who recalled seeing government workers, sent with eviction notices and appraisals, turn back in defeat. Fearing the potential tearing of their suits or the stinging cuts from attempts to maneuver past, they would ultimately be obstructed by the thick and dense matts of camelthorn growing across the arid border between the Black and the White City. Other neighbors would add that it was only proper for the camelthorn to behave this way, especially since the sharp needles of the plant were believed to prick the evil eye and protect from malicious action—including the planned demolition of the settlement and the further displacement of the community. While weediness has often been depicted as an index of disinvestment and neglect which forecloses a potential future, in the Black City qualities of apparent ecological hostility have instead allowed the marginal and makeshift population to exist and survive without surveillance or intrusion.[45]

There is something anarchic about the way that plants impede the movements of the rest of the city—dogwalkers from the residential developments rarely venture beyond the concrete pavements, while tourists assume that the vegetal disinvestment signals potential danger. Overpopulated with camelthorn clusters and devoid of high rises and glitzy developments, the Black City has retained distinct and generative qualities of marginality—species and places deemed too ungovernable to alter via governmental mandates or claims of illegal practice. Due to the difficulty of the landscape and its vegetation, fruitful guerrilla-gardening has emerged that resists the "gleaming spectacle of education and entertainment."[46] These gardens lack the glimmering technological apparatuses of contemporary horticulture, as well as the orderly, symbiotic relationships found in municipal parklands and gardens. They thwart the sovereign power of the state and the everyday contingencies of capitalism. The allotments of the Black City have become sites of unsupervised world-making, where both weeds and internally displaced communities articulate a different vision of settlement and inhabitation.

The weeds of the Black City foster a freedom that can only be found through remaining on the margins. One will rarely encounter a resident in the settlement who pays property tax. Water and electricity are regularly siphoned from networks, cloaked by the inaccessible flora and the overflow of illegal arrangements. Feral species culled elsewhere in the city have been offered sanctuaries to hide in, with plentiful resources to exploit. All three of these instances constitute distinct articulations of freedom for populations routinely targeted for destruction or removal. In many ways, the existence of the noxious weeds and the usual reluctance of the residents to remove them forges a potent form of community activism that rubs against mainstream greening initiatives. The antagonism and oppositional ethos of this world at war uses weediness as a point of entry for resistant civics shared with more-than-human coinhabitants. Weed-filled zones provide ideal conditions for displaced people who seek to maintain hard-fought-for autonomy. This means not having to rely on government handouts, which come with strict obligations to report and participate in bureaucratic processes. The autonomous people of the Black City reject mainstream narratives of refugee subjecthood, rooted in militarization, longing, nationalism, and melancholia. They seek to remain in a space where they feel part of a meaningful, multispecies, sociohistorical milieu.

Mediating the Margins

Months after my wound had finally transformed into a collection of keloid scars, I returned to the spot where I had encountered the stray dog. Instead of the cool sweep of the Khazri, the warm southern wind of the Gilavar glided across the allotments. The canine culprit, who I remembered via the impression of teeth, searing pain, and nightly fevers, as well as a slight limp, was curled up, slumbering.

Behind him, I could see the outline of the rest of the pack, the clustering of rough and ready houses with metal sheets reflecting the afternoon sun, as well as the sun-like flares of the refinery. In that moment, I thought back to the Russian word for a yard-dog, dvornyaga. This word is associated with both dvorovoi—a slavic spirit of the courtyard, mischievous and vicious toward livestock—and domovoi, a protective house spirit who tends to the well-being of kin. The same word, a holding place for tense contradiction, for an indeterminate spirit. It bares yellowed fangs of both justice and violence. Between the Black and White City, a thorny tangle, with flea-ridden fur and divisions, has emerged—inherited from rampant extraction and capitalist violence. Here multispecies mediations are unearthing opportunities with conditional forms of fracture, breakage, and ruination, rather than in visions of eternal gardens and biodiverse havens. The stray dog who bit me was a keeper of the extractive zone—a figure essential to the maintenance of a place that holds its own forms of life and life making—yet also devalued and scheduled for destruction under the guise of utopian environmental city planning. Here he was, mediating between the past and future, the industrial and the natural, the just and the unjust. I let the dvornyaga rest.

Notes

1. Whitehead, *Modes of Thought*, 116.
2. One may liken a state of attentiveness to David Bloor's notion of a gut feeling, which can be both an "indicator and expression of a moral response." "Epistemic Grace," 274.
3. Within this model, injustice becomes a case of poor distribution, which can be corrected via state action and recognition. Demands of justice stay within the confines of state institutions and organizations, which appear seemingly impartial, rational, and reasonable. See Low and Gleeson, *Justice, Society and Nature*.
4. Bauman, *Postmodern Ethics*, 42.
5. Low and Gleeson, *Justice, Society and Nature*, 38.

6   Haraway, *The Companion Species Manifesto*.
7   Freccero, "Figural Historiography," 50.
8   Weaver, "Pit Bull Promises," 352.
9   Gilligan et al., *Mapping the Moral Domain*, 8.
10 Lyons, "Two Perspectives," 135.
11 Latimer, "Being Alongside," 98.
12 Celermajer et al., "Justice through a Multispecies Lens."
13 Haraway, *Simians, Cyborgs and Women*.
14 Adams and Hutton, "People, Parks, and Poverty"; Brockington and Igoe, "Eviction for Conservation."
15 Latour, *War of the Worlds*.
16 Basl, "Restitutive Restoration."
17 It is worth considering the processes which have led to the subsumption of displaced individuals into an acronym. The "IDP" category in effect cleaves displacement spatially. While refugees are considered those who have crossed the territorial lines of the nation, internally displaced people enforce the integrity of the sovereign state. The category acts to reify the notion of a legible claim to territory by Azerbaijan through conceptualising the population, and the contested space of Nagorno-Karabakh, as still within the bounds of the country. According to this logic, the individuals resettled from the region have never been outside of Azerbaijan, thus they can only be internally displaced, denying Armenia's claims to possession, as well as self-determination.
18 Latour, "An Inquiry into Modes of Existence."
19 Tsing, "Threat to Holocene Resurgence," 52.
20 Blasted landscapes, Tsing writes, have experienced multiple environmental disasters. These disasters may include damage stemming from capitalist accumulation, industrial agriculture, militarization, or resource extraction. See Tsing, *Mushroom at the End of the World*. On scales of mattering, see Oppermann, "Scale of the Anthropocene."
21 Stoetzer, "Ruderal Ecologies," 309.
22 This brings to mind Stengers's definition of diplomacy as a thinking "through the middle" of established polemical positions, that enables new possibilities and the situated coordination of action. See Stengers, "Ecology of Practices."
23 Akin to what Rousell terms "little justices," small-scale acts of justice may entail a movement, an idea, or a ritual which may result in a creative endeavour (or not). It becomes what Washick et al. call the precondition for any deliberate attempt. See Rousell, "Doing Little Justices"; Washick et al., "Politics That Matter."
24 Josephson et al., *Environmental History of Russia*.
25 Harris-Brandts and Gogishvili, "Architectural Rumors"; Fikret, Huseyn, and Mammad, "Planning Sustainable Social Infrastructure."
26 All interlocutors have been anonymized in the piece.
27 One may consider this a distinctly post-Independence perspective, orienting itself against Soviet ecology which emphasized internal dynamics, contradictory changes, and instability. See, for example, Sukachev's dialectical integrative approach in Foster, "Late Soviet Ecology."

28 Absheron is a peninsula in Azerbaijan comprised of three districts: two are urban (Baku and Sumqayit) and one is semirural, composed of a number of villages (Absheron Rayon). Some villages, like Gala, were first settled over five thousand years ago. The village of Nardaran mentioned here evokes for many urban residents a pastoral ideal of sandstone houses, Soviet-built greenhouses, and a supposedly two-hundred-year-old olive tree. However, Nardaran is also a heavily policed settlement. After a 2015 crackdown on suspected Islamist radicals, the government reopened a long-closed police station and set up a local branch of the Ministry of National Security. For several years, a security checkpoint controlled all traffic in and out of the settlement, and residents were subject to frequent raids by the authorities. See Anonymous, "Nardaran"; Aliyeva, "Role of Landscape in the Formation of the Absheron Reserves."

29 The government has not revealed the prices of the foreign trees, which are thought to cost several hundred dollars, in addition to travel expenses and customs duties. In contrast, agronomists say a local olive tree would have cost the government thirty to forty manats, or around fifty US dollars. See Kazimova, "Azerbaijan's Extravagant Olive Trees."

30 Kazimova, "Azerbaijan's Extravagant Olive Trees."

31 *Bakinets* is a local term denoting people native to Baku and city dwellers with multigenerational ties to the capital, as opposed to the internally displaced population from the regions (the rural communes of Nagorno-Karabakh).

32 Åsbrink, "Nobels in Baku."

33 Åsbrink, "Ludvig Nobel Enters the Fight for Oil."

34 Ibrahimov, "Craftsmen's Quarter."

35 Akhundov, "Legacy of the Oil Barons," 10.

36 Voskoboynikov, "Mineralogicheskoye Opisaniye Poluostrova Apsherona."

37 Bookchin, *Ecology and Revolutionary Thought*, 89.

38 De Oliveira Andreotti et al., "Mapping Interpretations of Decolonization," 22.

39 Hospicing can be thought of as staying within modernity's frames in order to learn from it and clear space for something new. While recognizing the violence wielded by Enlightenment thinking, both epistemologically and materially, hospicing refutes a wholesale rejection of modernity and its philosophical tenets. Instead of shifting quickly to a new system, hospicing offers to modernity a form of palliative care. In my opinion, this sort of tactful approach proves generative when examining other contested, or even contestable, epistemological frames. See de Oliveira Andreotti et al., "Mapping Interpretations of Decolonization."

40 See Serpell, *In the Company of Animals*, 140–41; Kohn, "How Dogs Dream"; Schaffer, *One Nation under Dog*.

41 Gabardi, *Next Social Contract*, 153.

42 It should be noted that protocols are coproduced—a contract between two or more parties. They do not seek to insert humans into deterministic causal relations so that the beliefs, opinions, passions, and activity of humans are revealed to be the results of naturalistic forces. Instead, protocols are reflective of what Jane Bennett refers to as the "interfolding network of humanity and nonhumanity"

which creates shares of agency and structures relationships. See Bennett, *Vibrant Matter*, 31; Latour, "On Interobjectivity."

43 The global dump is offered up as an antisocial place where everything exists in pure abstraction: elements, people, and things all unshapen and unable to connect, or even relate, to one another. Indeed, the very material context is, in the words of Marder, "too volatile to support anything," let alone a social milieu. "Being Dumped," 192.
44 Murphy, "Alterlife and Decolonial Chemical Relations," 497.
45 See Hetherington, *Infrastructure, Environment, and Life in the Anthropocene*.
46 Myers, "From Edenic Apocalypse to Gardens against Eden," 118.

Closing
Th S xth M ss Ext nct n

Craig Santos Perez

"An m ls surro nded our anc st rs. An m ls wer th ir fo d, cl th s, adv rs ries, c mp nions, jok s, and th ir g ds…In th s age of m ss ext nct n and th ind str al zat on of l fe, it is h rd to touch th sk n of th s l ng and de p c mp n onsh p. N w we surro nd th an m ls and cr wd th m fr m th ir hom s…An mal ty and hum n ty ar one, xpr ss ons of th pl n t's br ll ant inv nt ven ss, and yet th an m ls ar leav ng th wor d and n t ret rn ng."

--Al son H wth rne Dem ng
Zool g es: On An m ls and th Hum n Sp r t (2 14)

Afterword
Fugitive Jurisdictions

Karin Bolender, Sophie Chao, and Eben Kirksey

WE MUST BELIEVE that words can do justice, right? If not, we wouldn't be here reading them, writing them, or chanting them in the streets. Certainly, words can do justice, for humans, in some ways. Our voices and articulations connect the fleeting fleshes we live in and among, and they stir our collective capacities for remembering, meaningful exchange, and action—for loving and caring, for resisting and rejecting. Our languages allow us to say the names of who and what we love and lose, bodies and places we hold close in lives and afterlives, and those we hope to carry into the future. They give us means to call out what must change and to "name the masters of broken earths," as Kathryn Yusoff demands.[1] In paradoxical ways, words can even hold spaces in which to say names we don't yet know and may never reckon.

Meanwhile the power of words to do justice can be a meager one, when pitted against the grinding forces of extractive industries and the endless horizons of toxic legacies. In seeking to describe how to do justice for frayed forests, overburdened oceans and waterways, for the casualties of ravenous plantationscapes and other seeming wastelands of wracked earthly ecologies—

and for the lives they may yet sustain—we rattle against the limits of our languages in other ways, too. Still we muster and polish our best syntaxes and vocabularies, as self-appointed advocates for victims of inequity in ecologies we speak with and for. All the while, humble attention reminds us that the communal ecologies for which we speak, and so the justices we seek for them, always overflow the margins of what we can say or know.

The contributions to this volume invite us to enter realms of others who recognize and move within their own codes of just relations, encoded in wild, wordless fluctuations of bites and bows, scents or saltiness, soundscape or temperature gradients, or even by trails of sensed significance we can't perceive—never mind translate. The roadside corpse of a young bull half-eaten by wild leopards exerts spectral and taxing demands on a woman who cannot escape his haunting presence in the central Himalayas. An endangered Micronesian kingfisher meets a fellow native of Guam with resounding, avian silence—and an invitation to rewrite extinction stories. The territory of the Colombian Amazon is one consequential actor, among many others, that plays a role in conflicts over intergenerational and intercultural justice. In the occupied nuclear zones of Indian Country, faith in the capacity of sacred lands to heal themselves with the aid of Native caretakers persists, despite dysfunctional kinships and literal toxicities that extend into the future in ways that humans can barely reckon. *The Promise of Multispecies Justice* points toward possibilities for dismantling hegemonic dictums that flatten a multiplicity of worlds and words in motion and open up opportunities for reengaging with what *just is*.

With linguistic limits in view, these essays venture into realms beyond dominant sites and scales of western colonial justice, following fainter paths and feral figures into brimming borderlands of ethnographic theory and narrative. Authors have tracked thunderous ghosts, wily weeds and strays, momentous viral particles, fading birdsongs, perpetual toxic traces, and evermore-visible marks of erasures—marking new possible trails for the pursuit of justice beyond human domains. At the same time, *The Promise of Multispecies Justice* recognizes that no one form of writing, discipline, or perspective can hope to hold all the prickly intersectional knots, possible alliances, and affective pulses at play in projects of justice-making within multispecies worlds.

In a mode that M. L. Clark generatively calls "inscribing the gap," this volume explores different ways that would-be justice-makers in multivocal worlds might broaden our senses of jurisdiction—as in who claims authority to "speak the laws" of the land. Haunting absences and presences reverberate in the open spaces of Craig Santos Perez's "Th S xth M ss Ext nct n," while Clark's speculative fictions offer imaginative propositions for unforeseen and otherworldly acts of justice-making. These experimental forays remind readers that no one authority can ever say what is just for all lives in any given situation.

Let's take a moment to consider how the bounds of jurisdiction might shift in multispecies worldings, where differently situated lives grasp and articulate "laws of the land" through wildly diverse ecosocial and sensorial dimensions. Such jurisdictions by nature must overflow the limits of legislative precedents and jurisprudence done by writs or citations. Outside domains of legal ledgers and binding promissory contracts, myriad interwoven forms of life feel, taste, smell, or otherwise perceive just or unjust relations through shifting pulses and flows, chemical sensations, and cyclical exchanges.

Even though some lives may perceive what we call justice in ways that do not translate, we can make spaces in which to recognize them, refusing to disqualify them just because they do not participate in accepted modes of knowing, naming, and juridical proceeding.[2] In approaching the limits of languages' dominions, we venture to ask: Can unruly poetic acts and artistic interventions cultivate spaces at the frayed edges and intersections of multispecies worlds, where wild patches of justice might grow for a time, perhaps even long enough to take root in unexpected places? Is it possible to coproduce seeds of justice and fling them forth through webs of interlacing agencies? Can we recognize species of justice (and/or injustice) that flourish beyond the grasp of sense and syntax?

Questions related to other-than-human un/knowability, language, and creative inquiry brought a regular gang of scholars and artists to gather weekly in a virtual space for a workshop on multispecies justice in the spring and summer of 2020, as the global pandemic wore on.[3] On one midsummer occasion, the workshop welcomed some guest artists and their special companions, who offered speculative proposals for reckoning with quandaries of just relations in surprising ways. Joining first from the crack of dawn in Taipei, bioartist and provocateur Kuang-Yi Ku shared a number of projects that suggest speculative solutions to conflicts around endangered species conservation and Traditional Chinese Medicine (TCM).[4] Ku's *Tiger Penis Project* mischievously blurs categories of technoscientific enterprise and artful social practice to dig into deeply held values at the crux of human-wildlife conflict. Proposing (half seriously) to bioengineer the tissues of the imperiled big cat's genitals (and so their unique health benefits in TCM) in a laboratory, the *Tiger Penis Project* models a mode of wily prosthetic fabulation that pokes inward from the edges toward the core beliefs at play in seemingly impassable conflicts.

Just as the end of that session drew near, Jan-Maarten Luursema appeared suddenly on the screen from deep in a dark Dutch night. With a brief introduction to his laconic companion, he transported the assembled crew into the realm of an improbable project that radically enacts collaborative interlocution with earthly lifeforms who seem alien to human ways of moving, knowing, and individual bodying. And so began a provocative exchange across material-semiotic weavings of mind, matter, and meaning on questions of beyond-human justice with a slime mold (*Physarum polycephalum*) known as Andi.

Captured some years ago by the slime mold's mysterious ways, Jan-Maarten Luursema began to wonder what this curious earthly creature might feel or think or even have to say. So, he developed a mesh of technical apparatuses to enable Andi to spell out words. Through a Twitter account, @slimemoldAndi, the artist invites humans to pose linguistic questions. Andi then responds to these verbal queries by growing slowly across a Ouija-like alphabetic field set into a medium of agar in a petri dish, exploring the space in uniquely streaming plasmodial ways while grazing on a preferred diet of oats. Once the slime mold's growth cycle in this medium has run its course (usually after a few days to a week, and often in duet with a blossoming black mold), Jan-Maarten inputs the chain of letters that Andi's movement has linked together into Word's spellchecker. This process generates Andi's response to the question posed, which is then offered back to the questioners to interpret (see figure C.1).[5]

Much fascinating theoretical, scientific, and creative experimental work unfolds these days in the company of slime molds, especially *Physarum polycephalum,* the domestic workhorse of the myxomycetes. While Aimee Bahng draws on the world-making wisdoms of Octavia Butler to probe the radical social possibilities posed by plasmodial lifeways (and to warn against their cooptation by corporate enterprises), others revel in the implications of their unfathomable, many-headed intelligences.[6] Steven Shaviro regards slime molds as an "extended mind" with cognition that involves the coupling of the organism, external resources, and the biochemical memory traces they inscribe into their environments. Even though slime molds do not have brains, or even mouths, Shaviro argues, "*Physarum* feels, and ponders, and decides. It acts in ways that are not always stereotypical, but at least to some degree spontaneous." He concludes, "There is certainly sentience here."[7]

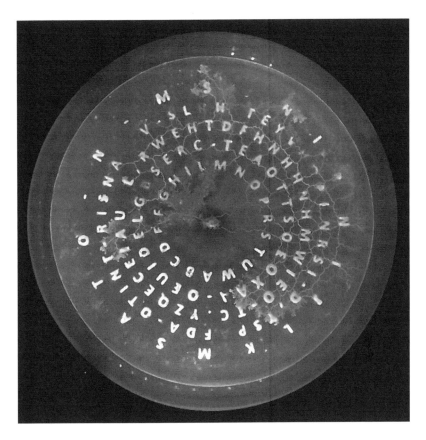

C.1 Slime mold Andi engaged in conversation. Photograph by Jan-Maarten Luursema.

But the limelight cast on the many wondrous capacities of slime molds in recent years is not really what lured our multispecies justice workshop to linger at the lively edges of their worldings. It was in fact on another lockdown Zoom call one early summer morning—a special gathering of the international Slime Mould Collective hosted by artist Heather Barnett—where one of us (Karin) was first captivated by Jan-Maarten Luursema's suggestion that slime molds *sing*: he described how one might walk through a forest unknowing, all the while immersed in subaudible choruses of wild, unseen slime molds singing their lives. And this in turn evokes, echoes, and amplifies the vast realms of lively exchange to which dominant human senses, ontologies, and so too jurisdictions are utterly oblivious.

From the underground mycorrhizal networks that help forests distribute resources to the more overt gestures, tussles, or just-hard-enough bites and body checks through which other mammals balance perceived rights and wrongs, most jurisdictions fail to recognize embodied orders of just relations that fall outside and/or do not register in linguistic arguments. Given how much of this worldly adjudication transpires in wordless ways, it seems that questions of multispecies justice must carefully consider how other species experience justices beyond the long arm of the laws of languages—even as we remain bound by their modes of connection and coercion. As we seek to meaningfully expand registers of justice to include more lives, language may well present a border worth poking and prodding for promising portals, beyond which different jurisdictions blaze up exchanges of justice in different registers altogether.

But as we consider the implications of thinking differently with slime molds and other companion species that inhabit shared multispecies worlds, one key question looms large as we try to ground broadening ethical and political concerns: whose lives matter? Several contributors to *The Promise of Multispecies Justice* mobilize the cosmopolitical ideas of Isabelle Stengers to wrangle with the ongoing antagonism and conflicts that persist in life-and-death struggles in multispecies contact zones. Stengers's notion of "reciprocal capture" offers inspiration as we dwell on the question: whose lives matter in multispecies worlds? She describes how beings sometimes fold one another into enduring relations, or symbiotic agreements, when transformative encounters produce new, entangled modes of coexistence.[8] In other words, the lives of beings matter to each other in distinct and complicated ways, when they become entangled in shared cosmopolitical worlds.

As we contemplate an ethics of inclusion, it is critical to also follow Eva Haifa Giraud to recognize how exclusions as "entities, practices, and ways of being are *foreclosed* when other entangled realities are materialized."[9] Exclusions are made to clear space for loved ones or to contest destructive relations, intuitions, and processes. In a deliberate mode of reckoning with dynamics of inclusion and exclusion at smaller scales, artistic interventions can offer unexpected spaces and encounters through which to consider specific frictions, foreclosures, and exclusions that come with any attempt to keep a cherished form of life in the world.[10] Killing sometimes becomes necessary when one makes commitments to care for others within complex ecosystems. Taking a stand for particular bodies or communities often requires standing against other organisms that become enemies or hostile forces in human social worlds.[11] In their careful tracing of invisible or taken-for-granted frames and boundaries, creative and activist interventions can act as reminders of the limits of any one judge or author to say what justice might feel like for everyone involved in a given milieu.

By making space for a slime mold, and even entertaining the possibility of conversing with Andi's apparatus, we tried to honor Isabelle Stengers's suggestion "that collective thinking has to proceed 'in the presence of' those who would otherwise be likely to be disqualified as having idiotically nothing to propose, hindering the emergent 'common account.'"[12] Whatever meanings we make of them, engaging with otherworldly interlocutors is an act of radical inclusion and rightful recognition of interspecies limits and gaps. More than that, minding these gaps might help to "unthink mastery" and create modes of meaningful, if playfully unruly, exchange.[13]

If justice ultimately aims to care for and cultivate what a given collective loves, then intimacy, creativity, and play are as vital to the work as antagonism, defiance, and rejection. In the communal doings of other social mammals, like whales and canines, play holds significant training in the sorts of acts and exchanges that right wrongs and seek to keep the peace in a pack or pod. Play is a mode in which to test limits, to practice common forms of life, and to probe the edges of them, experimenting and innovating. Whether as a rambunctious wrestling session or an artistic experiment with unfamiliar lifeforms, play is practice for the necessities of living, thinking, moving, caring, standing ground, and feeling out the edges of shared worlds. Play helps sustain and nurture relations within collectives, as each member must learn how hard to bite and how to respond to a cry. Arenas for play can serve as learning grounds for recognizing when to rise up and fight, or when—and how far and fast—to run.

Sometimes a body must flee from places where one has done all the fighting for justice that one can do. At times, flight from suffocating forces—toward promises of freedoms and recognition that can only live and breathe beyond them—is the only way to survive. Flight, then, is means to finding justice, even though fights for justice are easier to see and celebrate. Whether escaping structures that reinforce generations of bodily harm or breaking out of oppressive jurisdictions, the pursuit of justice can demand breaches into fullblown fugitive territories, where unrecognized sovereignties and unexpected allies find each other and forge new just relations. In recognizing the forms of life that are flourishing in these fugitive jurisdictions, we are indebted to the poetics and "fugitive aesthetics" of contemporary Black feminist, Indigenous, and decolonial art, activism, and scholarship, which enact modes of creative resistance that overflow and resist colonial confinement and containment.[14]

Fugitivity is a strategy long lived and practiced in Black, queer, Latin American, Asian, and Indigenous lives and writings. It is a response to falsely imposed structures that violently reinforce the exclusion of some human lives from freedoms, dignities, and equities that others take as theirs by right. Fugitives borne in words and actions resist, mark, and defy these impositions in their ongoing struggle against social death.[15] We must not sever theories of fugitivity from the long line of radical and Black intellectuals who trace their thinking directly to Frederick Douglass and other people who have sought to escape the pervasive violence of chattel slavery. Black fugitivity persists in the afterlives of slavery as "a desire for and a spirit of escape and transgression of the proper and the proposed," in the words of Fred Moten.[16]

In writing about the undercommons—a figurative space inhabited by Black, Indigenous, queer, and poor peoples—Jack Halberstam suggests that shared desires unite fugitives who "want to take apart, dismantle, tear down the structure that, right now, limits our ability to find each other."[17] As many who live in the undercommons use opacity and invisibility as strategies of survival, Karma Chávez invites fugitives who find refuge to embrace queer politics of visibility that insists both "Aqui, estamos, y no nos vamos" and "We're here, we're queer, get used to it!"[18] In a world facing system collapse, creating new refuges for oddkin, human and not, offers generative spaces where unruly forms of life might sustain rich cultural and biological diversity.[19]

Fugitive lives matter. Authors in this collection have followed the precarious lives of displaced peoples, plants, and animals to places where they have created their own refuges. Stray dogs and tenacious camelthorn in the Black City of Baku remind visiting ethnographers and city planners that they are trespassing into fugitive jurisdictions. Elizabeth Lara's work shows us that it is even possible to create refuges within spaces that are inherently hostile to plants and people—like the prison—even while harboring hopes for new forms of multispecies justice that might grow beyond and displace conventional modes of discipline and punishment. Slipping through rusty barbed-wire fences and cracks in crumbling concrete, justice here and elsewhere defies the bastions of institutionalized, western-colonial jurisprudence and associated jurisdictions.

Fugitive spaces elude conventional definitions and dictions of justice; yet they might still be inviting to those who pursue multispecies justice and bust through borders that block some lives' paths to flourishing. For those of us who happen to read and write and speak, escaping the bounds of the languages that hold us is not a plausible path to move beyond the false divisions and inequitable distributions that words and writs too often enforce.

Yet by bending and twisting the chain-linked grammars and logics that divide and exclude some bodies, we might still open avenues into vital communal spaces held together in ecological webs of matter and meaning. Catastrophic times call for catachrestic idioms that capture the contradictions and dilemmas of a broken world through languages patchy, partial, and out of place.[20] It is time to make new spaces for lively grammars of animacy that pave the way to openings after words, untethered from the hegemony of human languages.[21]

By creating disruptive openings for others to join the conversation, creative and speculative projects merge with centrifugal ethnographic works as forays into the elusive fugitive jurisdictions of multispecies worlds. Taken together, the texts in this collection all ask deeper ecological questions of what, how, and by whom laws and languages of any land are comprised. Unsung fugitives who escape capture (and maybe even notice) within recognized jurisdictions offer discreet getaways that may lead, if by faint scent-marks and inaudible songs, into untamed territories. At the frayed edges, perimeters, and in the spaces-between human justices, we just might recognize previously unheard voices of myriad insects, plants, waterways, forests, spores, and other outlaws, singing their own corridos—old songs of love and loss, daring escapes and even little justices: recalled, remembered, lived in and longed for.

Notes

1. Yusoff, *Billion Black Anthropocenes or None*, 2. See also Gray and Sheikh, "Wretched Earth."
2. As Astrida Neimanis and Sria Chatterjee write, "But even as—or perhaps because—some lifeworlds are unassimilable or inscrutable, we are still faced with the task of including them in the project of multispecies justice: we need to both know them and resist mastering them through knowledge. The question is: how?" Celermajer et al., "Justice through a Multispecies Lens," 493.
3. Regular guests included authors and editors in this collection, along with anthropologist Michele Lobo, animation artist Gina Moore, and other visitors, all of whom contributed vital insights to conversations around questions of multispecies justice.
4. For more of Kuang-Yi Ku's projects online, see Ku, "Kuang-Yi Ku."
5. On this occasion, the question we posed to the slime mold's conversation apparatus was this: "Andi, what might multispecies justice feel or smell or taste or sound like?" Andi's elusive response invites further pondering (@slimemoldAndi): "Minnesota lithesome watersheds, gangrenous. Eibar toes intrico also am. K. Tsar

ditto wiring pup though nine beehives," Twitter, August 10, 5:10 p.m. https://twitter.com/slimemoldAndi/status/1292931353688903686/.
6   See Bahng, "Plasmodial Improprieties."
7   Shaviro, *Discognition*, 197–214.
8   Stengers, *Cosmopolitics I*, 35–36.
9   Giraud, *What Comes after Entanglement?*, 2.
10 Giraud, *What Comes after Entanglement?*, 2.
11 Kirksey, *Emergent Ecologies*, 218.
12 Stengers, "Cosmopolitical Proposal," 1002.
13 Singh, *Unthinking Mastery*. This conversation with the slime mold also seeks to cultivate what Astrida Neimanis and Sria Chatterjee call "intimacy without mastery." Celermajer et al., "Justice through a Multispecies Lens," 491–97. On the political, ethical, and affective matterings of art and visual culture for multispecies justice in an age of climate breakdown, see also Agarwal, "Alien Waters," and Broglio, "Multispecies Futures through Art."
14 Martineau and Ritskes, "Fugitive Indigeneity," iv. See also Biddle, *Remote Avant-Garde*; Todd, "Indigenizing the Anthropocene"; and Gumbs, *Spill*.
15 von Gleich, "African American Narratives of Captivity and Fugitivity."
16 Moten, *Stolen Life*, 134. See also Hartman, *Lose Your Mother*; Morgan, "Accounting for the 'Most Excruciating Torment'"; and Trouillot, "Culture on the Edges."
17 Halberstam, "The Wild Beyond," 6.
18 Chávez, "Queer Politics of Fugitivity," 68.
19 Haraway, *Staying with the Trouble*, 147.
20 Tsing, "Catachresis for the Anthropocene," 2–3.
21 De la Cadena and Blaser, *World of Many Worlds*, 4; Kimmerer, *Braiding Sweetgrass*, 57–58.

Bibliography

Abraham, Delna, and Ojaswi Rao. "84% Dead in India's Cow-Related Violence." *Hinduism Times*, June 28, 2014. www.archive.indiaspend.com/cover-story/86-dead-in-cow-related-violence-since-2010-are-muslim-97-attacks-after-2014-2014.

Abu-Jamal, Mumia, and Angela Y. Davis. *Jailhouse Lawyers: Prisoners Defending Prisoners v. the USA*. San Francisco: City Lights Publishers, 2009.

Adams, William M., and Jon Hutton. "People, Parks, and Poverty: Political Ecology and Biodiversity Conservation." *Conservation and Society* 5, no. 2 (2007): 147–83.

Adcock, Cassie. "Sacred Cows and Secular History: Cow Protection Debates in Colonial North India." *Comparative Studies of South Asia, Africa, and the Middle East* 30, no. 2 (2010): 297–311.

Admin. "Rethinking the Apocalypse: An Indigenous Anti-Futurist Manifesto." *Indigenous Action*, March 19, 2020. www.indigenousaction.org/rethinking-the-apocalypse-an-indigenous-anti-futurist-manifesto.

Administrator. "Mosqueda v Pilipino Banana Growers and Exporters Association." *Lex Animo*, August 16, 2016. www.lexanimo.com/2021/03/12/mosqueda-v-pilipino-banana-growers-and-exporters-association/.

"Aerial Spraying 'Safest, Most Effective.'" *SunStar Newspaper*, March 23, 2018. www.sunstar.com.ph/article/185107.

Agard-Jones, Vanessa. "Bodies in the System." *Small Axe* 17, no. 3 (2013): 182–92.

Agard-Jones, Vanessa. "Ep. #35 – Vanessa Agard-Jones." *Cultures of Energy* (podcast), September 29, 2016. Accessed January 15, 2022. https://cenhs.libsyn.com/ep-35-vanessa-agard-jones.

Agarwal, Ravi. "Alien Waters." In *The Routledge Companion to Contemporary Art, Visual Culture, and Climate Change*, edited by T. J. Demos, Emily Eliza Scott, and Subhankar Banerjee, 365–75. New York: Routledge, 2021.

Ahmed, Sara. *The Promise of Happiness*. Durham, NC: Duke University Press, 2010.

Akhundov, Fuad. "Legacy of the Oil Barons: The 'Greening of Baku' by the Nobel Brothers." *Azerbaijan International* 2, no. 3 (1994): 10–11.

Alaimo, Stacey. *Bodily Natures: Science, Environment, and the Material Self*. Bloomington: Indiana University Press, 2010.

Aliyeva, Rahiba. "The Role of Landscape in the Formation of the Absheron Reserves." *International Journal of Advanced and Applied Sciences* 5, no. 6 (2018): 35–44.

Allen, Barbara L. *Uneasy Alchemy: Citizens and Experts in Louisiana's Chemical Corridor Disputes*. Cambridge, MA: MIT Press, 2003.

Anderson, Bridget. "The Politics of Pests: Immigration and the Invasive Other." *Social Research* 84, no. 1 (2017): 7–28.

Anderson, Elizabeth S. "What Is the Point of Equality?" In *Theories of Justice*, edited by Alejandra Mancilla and Tom Campbell, 133–83. New York: Routledge, 2017.

Anonymous. "Nardaran: Still a Village Apart." *Chai Khana*, December 17, 2017. https://chaikhana.media/en/stories/607/nardaran-still-a-village-apart.

Appadurai, Arjun. "Introduction: Place and Voice in Anthropological Theory." *Cultural Anthropology* 3, no. 1 (1988): 16–20.

Arendt, Hannah. *Between Past and Future: Eight Exercises in Political Thought*. New York: Penguin, 2006.

Ariza, Eduardo, María C. Ramírez, and Leonardo Vega. *Atlas Cultural de la Amazonia Colombiana: La Construcción del Territorio en el Siglo XX*. Bogotá: ICANH, 1998.

Armenteras, Dolors. "¿En qué quedó el medio ambiente? Un gran daño a nuestros bosques." *Razonpublica*, December 17, 2018. razonpublica.com/en-que-quedo-el-medio-ambiente-un-gran-dano-a-nuestros-bosques/.

Arquiza, Yasmin D. "Davao City Govt, Farmers Push Ban on Aerial Pesticide." *GMA News*, October 13, 2008. www.gmanetwork.com/news/news/specialreports/126671/davao-city-govt-farmers-push-ban-on-aerial-pesticide-spraying/story.

Åsbrink, Brita. "Ludvig Nobel Enters Fight for Oil." *Brothers Nobel*, August 15, 2011. www.branobelhistory.com/themes/the-nobel-brothers/ludvig-nobel-enters-the-fight-for-oil/.

Åsbrink, Brita. "The Nobels in Baku: Swedes' Role in Baku's First Oil Boom." *Azerbaijan International* 10, no. 2 (2002): 56–59.

Bahng, Aimee. "Plasmodial Improprieties: Octavia E. Butler, Slime Molds, and Imagining a Femi-Queer Commons." In *Queer Feminist Science Studies: A Reader*, edited by Cyd Cipolla, Kristina Gupta, David A. Rubin, and Angela Willey, 310–25. Seattle: University of Washington Press, 2017.

Balane, Walter. "Banana Firms: Please Don't Ban Aerial Spraying." *MindaNews*, September 14, 2006. www.mindanews.com/c3-news/2006/09/banana-firms-please-dont-ban-aerial-spraying/.

Banerjee, Subhankar. "Resisting the War on Alaska's Arctic with Multispecies Justice." *Social Text Journal*, June 7, 2018. www.socialtextjournal.org/periscope_article/resisting-the-war-on-alaskas-arctic-with-multispecies-justice/.

Bantay Kinaiyahan. *People vs. Profit: A Briefer on the Ban Aerial Spraying Campaign in Davao*. Davao, Philippines: Interface Development Interventions Inc., 2006.

Baptiste, Kristie L. *Final Report: Hanford Tribal Stewardship*. Prepared for Nez Perce Tribe. Nez Perce Tribe Department of Natural Resource, Environmental Restoration and Waste Management, 2005. Accessed October 20, 2020. www2.clarku.edu/mtafund/prodlib/nez_perce/Hanford_Tribal_Stewardship.pdf.

Barad, Karen. *Meeting the Universe Halfway: Quantum Physics and the Entanglement of Matter and Meaning*. Durham, NC: Duke University Press, 2007.

Basl, John. "Restitutive Restoration: New Motivations for Ecological Restoration." *Environmental Ethics* 32, no. 2 (2010): 135–47.

Bateson, Gregory. *Steps to an Ecology of Mind: Collected Essays in Anthropology, Psychiatry, Evolution, and Epistemology*. Chicago: University of Chicago Press, 2000.

Bauman, Zygmunt. *Postmodern Ethics*. Oxford: Blackwell, 1993.

Bawaka Country, Sarah Wright, Sandie Suchet-Pearson, Kate Lloyd, Laklak Burarrwanga, Ritjilili Ganambarr, Merrkiyawuy Ganambarr-Stubbs, Banbapuy Ganambarr, Djawundil Maymuru, and Jill Sweeney. "Co-becoming Bawaka: Towards a Relational Understanding of Place/Space." *Progress in Human Geography* 40, no. 4 (2016): 455–75.

Baxter, Brian. "Ecological Justice and Justice as Impartiality." *Environmental Politics* 9, no. 3 (2000): 43–64.

Baxter, Brian. *Ecologism: An Introduction*. Washington, DC: Georgetown University Press, 1999.

Baxter, Brian. *A Theory of Ecological Justice*. New York: Routledge, 2004.

Bayly, Thomas. *The Royal Charter Granted unto Kings by God Himself*. London, 1656.

Beliso-de Jesús, Aisha. *Electric Santería: Racial and Sexual Assemblages of Transnational Religion*. New York: Columbia University Press, 2015.

Benjamin, Ruha. "Black Afterlives Matter: Cultivating Kinfulness as Reproductive Justice." In *Making Kin Not Population*, edited by Adele Clark and Donna J. Haraway, 41–66. Chicago: Prickly Paradigm Press, 2018.

Benjamin, Ruha, ed. *Captivating Technology: Race, Carceral Technoscience, and Liberatory Imagination in Everyday Life*. Durham, NC: Duke University Press, 2019.

Benjamin, Ruha. "Racial Fictions, Biological Facts: Expanding the Sociological Imagination through Speculative Methods." *Catalyst: Feminism, Theory, Technoscience* 2, no. 2 (2016): 1–28.

Benjamin, Ruha, and Eddie Glaude. "Reimagining Science and Technology." *African American Studies* (podcast), March 12, 2018. Accessed September 20, 2020. www.aas.princeton.edu/news/aas21-podcast-episode-12-reimagining-science-and-technology.

Bennett, Jane. *Vibrant Matter: A Political Ecology of Things*. Durham, NC: Duke University Press, 2010.

Bennett, Joshua. *Being Property Once Myself: Blackness and the End of Man*. Cambridge, MA: Belknap Press of Harvard University Press, 2020.

Besky, Sarah. "Exhaustion and Endurance in Sick Landscapes: Cheap Tea and the Work of Monoculture in the Dooars, India." In *How Nature Works: Rethinking Labor on a Troubled Planet*, edited by Sarah Besky and Alex Blanchette, 23–40. New Mexico: University of New Mexico Press, 2019.

Biddle, Jennifer Loureide. *Remote Avant-Garde: Aboriginal Art Under Occupation.* Durham, NC: Duke University Press, 2016.

Bishop, Claire. "Antagonism and Relational Aesthetics." *October* 110 (2004): 51–79.

Blanchette, Alex. "Herding Species: Biosecurity, Posthuman Labor, and the American Industrial Pig." *Cultural Anthropology* 30, no. 4 (2015): 640–69.

Blanchette, Alex. *Porkopolis: American Animality, Standardized Life, and the Factory Farm.* Durham, NC: Duke University Press, 2020.

Bloor, David. "Epistemic Grace: Antirelativism as Theology in Disguise." *Common Knowledge* 13, no. 2 (2007): 250–80.

Bohme, Susanna R. *Toxic Injustice: A Transnational History of Exposure and Struggle.* Berkeley: University of California Press, 2010.

Boisseron, Bénédicte. *Afro-Dog: Blackness and the Animal Question.* New York: Columbia University Press, 2018.

Bonhomme, Edna. "Troubling (Post)Colonial Histories of Medicine: Toward a Praxis of the Human." *Isis* 111, no. 4 (December 2020): 830–33.

Bookchin, Murray. *Ecology and Revolutionary Thought.* New York: Anarchos, 1979.

Botero, Roberto. "Señor Minambiente hablemos de deforestación en serio." *Cerosetenta*, October 5, 2020. www.cerosetenta.uniandes.edu.co/senor-minambiente-hablemos-de-deforestacion-en-serio/.

Boudia, Soraya, and Natalie Jas. *Powerless Science? Science and Politics in a Toxic World.* New York: Berghahn Books, 2014.

Bradshaw, Elizabeth A. "Tombstone Towns and Toxic Prisons: Prison Ecology and the Necessity of an Anti-Prison Environmental Movement." *Critical Criminology* 26, no. 3 (2018): 407–22.

Braverman, Irus. "Captive: Zoometric Operations in Gaza." *Public Culture* 29, no. 1 (2017): 191–215.

Brent, Keith J., and Derek W. Hollomon. *Fungicide Resistance in Plant Management: How Can It Be Managed?* Brussels: Fungicide Resistance Action Committee, Crop Life International, 2007.

Brockington, Daniel, and James Igoe. "Eviction for Conservation: A Global Overview." *Conservation and Society* 4, no. 3 (2006): 424–70.

Broglio, Ron. "Multispecies Futures through Art." In *The Routledge Companion to Contemporary Art, Visual Culture, and Climate Change*, edited by T. J. Demos, Emily Eliza Scott, and Subhankar Banerjee, 342–52. New York: Routledge, 2021.

Brooks, Dale L. "Animal Rights and Vertebrate Pest Control." *Proceedings of the Vertebrate Pest Conference* 13 (1988): 14–17.

Brooks, Thom. "Retribution." In *The Routledge Handbook of the Philosophy and Science of Punishment*, edited by Farah Focquaert, Elizabeth Shaw, and Bruce N. Waller, 18–25. New York: Routledge, 2020.

Brown, H. Claire. "How Corporations Buy—and Sell—Food Made with Prison Labor." *The Counter*, May 18, 2021. www.thecounter.org/how-corporations-buy-and-sell-food-made-with-prison-labor/.

Brown, Michelle. "Of Prisons, Gardens, and the Way Out." *Studies in Law, Politics, and Society* 64 (2014): 67–84.

Bubandt, Nils. "Anthropocene Uncanny: Nonsecular Approaches to Environmental Change." In *More Than Human: A Non-Secular Anthropocene—Spirits, Specters, and Other Nonhumans in a Time of Environmental Change*, edited by Nils Bubandt, 2–18. AURA Working Papers 3, 2018.

Bullard, Robert D. *Confronting Environmental Racism: Voices from the Grassroots*. Boston: South End Press, 1993.

Burkhalter, Eddie, Izzy Colón, Brendon Derr, Lazaro Gamio, Rebecca Griesbach, Ann Hinga Klein, Danya Issawi, K. B. Mensah, Derek M. Norman, Savannah Redl, Chloe Reynolds, Emily Schwing, Libby Seline, Rachel Sherman, Maura Turcotte, and Timothy Williams. "Incarcerated and Infected: How the Virus Tore through the U.S. Prison System." *New York Times*, April 10, 2021. www.nytimes.com/interactive/2021/04/10/us/covid-prison-outbreak.html.

Büscher, Bram. "The Nonhuman Turn: Critical Reflections on Alienation, Entanglement and Nature under Capitalism." *Dialogues in Human Geography*. Published ahead of print June 21, 2021. doi:10.1177/20438206211026200.

Buyandelger, Manduhai. "Asocial Memories, 'Poisonous Knowledge,' and Haunting in Mongolia." *Journal of the Royal Anthropological Institute* 25, no. 1 (2019): 66–82.

Cabot, Zayin. *Ecologies of Participation: Agents, Shamans, Mystics, and Diviners*. Lanham, MA: Lexington Books, 2018.

California Department of Corrections and Rehabilitation. "CDCR/CCHCS COVID-19 Employee Status." Accessed May 31, 2021. www.cdcr.ca.gov/covid19/cdcr-cchcs-covid-19-status/.

California Department of Corrections and Rehabilitation. "Monthly Total Population Report Archive." Office of Research. Accessed June 1, 2021. www.cdcr.ca.gov/research/monthly-total-population-report-archive-2019/.

California Department of Corrections and Rehabilitation. "Population COVID-19 Tracking." Accessed April 27, 2021. www.cdcr.ca.gov/covid19/population-status-tracking/.

California Department of Corrections and Rehabilitation. "Prison Closure Information." Prison Closures. Accessed January 11, 2022. www.cdcr.ca.gov/prison-closures.

California Department of Corrections and Rehabilitation. "Updates." Accessed October 2, 2020. www.cdcr.ca.gov/covid19/updates/.

California State Senate. "Senate Public Safety Committee, Wednesday, July 1st, 2020." Accessed July 7, 2020. www.senate.ca.gov/media/senate-public-safety-committee-20200701/video.

Celermajer, Danielle, Sria Chatterjee, Alasdair Cochrane, Stefanie Fishel, Astrida Neimanis, Anne O'Brien, Susan Reid, Krithika Srinivasan, David Schlosberg, and Anik Waldow. "Justice through a Multispecies Lens." *Contemporary Political Theory* 19 (2020): 475–512.

Celermajer, Danielle, David Schlosberg, Lauren Rickards, Makere Stewart-Harawira, Mathias Thaler, Petra Tschakert, Blanche Verlie, and Christine Winter. "Multispecies Justice: Theories, Challenges, and a Research Agenda for Environmental Politics." *Environmental Politics* 30, nos. 1–2 (2020): 119–40.

Centro Nacional de Memoria Histórica. *Narrativas de la Guerra a Través del Paisaje*. Bogotá: CNMH, 2018.
Centro Nacional de Memoria Histórica. *Petróleo, Coca, Despojo Territorial y Organización Social en Putumayo*. Bogotá: CNMH, 2015.
Chakrabarty, Dipesh. *The Crises of Civilization: Exploring Global and Planetary Histories*. New York: Oxford University Press, 2018.
Chakrabarty, Dipesh. *Provincializing Europe*. Princeton, NJ: Princeton University Press, 2009.
Chao, Sophie. "The Beetle or the Bug? Multispecies Politics in a West Papuan Oil Palm Plantation." *American Anthropologist* 123, no. 2 (2021): 476–89.
Chao, Sophie. "Can There Be Justice Here? Indigenous Perspectives from the West Papuan Oil Palm Frontier." *Borderlands* 20, no. 1 (2021): 11–48.
Chao, Sophie. "In the Shadow of the Palm: Dispersed Ontologies among Marind, West Papua." *Cultural Anthropology* 33, no. 4 (2018): 621–49.
Chao, Sophie. *In the Shadow of the Palms: More-Than-Human Becomings in West Papua*. Durham, NC: Duke University Press, 2022.
Chao, Sophie. "We Are (Not) Monkeys: Contested Cosmopolitical Symbols in West Papua." *American Ethnologist* 48, no. 3 (2021): 274–87.
Chávez, Karma R. "From Sanctuary to a Queer Politics of Fugitivity." *QED: A Journal in GLBTQ Worldmaking* 4, no. 2 (2017): 63–70.
Chigateri, Shraddha. "Negotiating the Sacred Cow: Cow Slaughter and the Regulation of Difference in India." In *Democracy, Religious Pluralism, and the Liberal Dilemma of Accommodation*, edited by Monica Mookherjee, 137–59. Dordrecht: Springer, 2011.
Choy, Timothy. "Distribution." *Fieldsights: Society for Cultural Anthropology*, January 21, 2016. www.culanth.org/fieldsights/distribution.
Choy, Timothy. *Ecologies of Comparison: An Ethnography of Endangerment in Hong Kong*. Durham, NC: Duke University Press, 2011.
Choy, Timothy, and Jerry Zee. "Condition—Suspension." *Cultural Anthropology* 30, no. 2 (2015): 210–23.
Cipriani, Joseph, Ashley Benz, Alanna Holmgren, Dana Kinter, Joseph McGarry, and Gabrielle Rufino. "A Systematic Review of the Effects of Horticultural Therapy on Persons with Mental Health Conditions." *Occupational Therapy in Mental Health* 33, no. 1 (2017): 47–69.
Clark, M. L. "To Catch All Sorts of Flying Things." *Clarkesworld*, September 2019.
Clarke, Adele E., and Susan L. Star. "The Social Worlds/Arenas/Discourse Framework as a Theory-Methods Package." In *The New Handbook of Science and Technology Studies*, edited by Edward J. Hackett, Olga Amsterdamska, Michael E. Lynch, and Judy Wajcman, 113–37. Cambridge, MA: MIT Press, 2008.
Clarke, Tracylee. "The Construction of Goshute Political Identity: Negotiation of Voice Regarding Nuclear Waste Policy Development." *Original Research* 2 (2017): 1–13.
Clifford, James. "Indigenous Articulations." In *Returns: Becoming Indigenous in the Twenty-first Century*. Cambridge, MA: Harvard University Press, 2013.
Coffey, Wallace, and Rebecca Tsosie. "Rethinking the Tribal Sovereignty Doctrine:

Cultural Sovereignty and the Collective Future of Indian Nations." *Stanford Law and Policy Review* 12 (2001): 191–221.

Cohen, Mathilde. "Animal Colonialism: The Case of Milk." *American Journal of International Law Unbound* 111 (2017): 267–71.

Cole, Luke W., and Sheila R. Foster. *From the Ground Up: Environmental Racism and the Rise of the Environmental Justice Movement*. New York: New York University Press, 2001.

Columbia Riverkeeper. *Hanford and the River*. Hood River, OR: Columbia Riverkeeper, n.d. Accessed October 25, 2020. www.columbiariverkeeper.org/sites/default/files/2011/10/hanford_and_the_river_final2.pdf.

Comaroff, Jean. "Invasive Aliens: The Late-Modern Politics of Species Being." *Social Research: An International Quarterly* 84, no. 1 (2017): 29–52.

Confederated Tribes of the Umatilla Indian Reservation. "A Brief History of CTUIR: Background Information on Our People." Cayuse—Umatilla—Walla Walla: Confederated Tribes of the Umatilla Indian Reservation. Accessed October 7, 2020. www.ctuir.org/about/brief-history-of-ctuir/.

Cooke, Stuart, and Peter Denney. *Transcultural Ecocriticism: Global, Romantic and Decolonial Perspectives*. London: Bloomsbury, 2021.

Cooper, David, and Joy Palmer. *Just Environments: Intergenerational, International and Inter-Species Issues*. New York: Routledge, 2005.

Cooper, Jessica. "Patience." *Fieldsights: Society for Cultural Anthropology*, June 22, 2018. www.culanth.org/fieldsights/patience.

Copeland, Marion. *Cockroach*. New York: Reaktion Books, 2003.

Copland, Ian. "Cows, Congress, and the Constitution: Jawaharlal Nehru and the Making of Article 48." *South Asia: Journal of South Asian Studies* 40, no. 4 (2017): 723–43.

Corpoamazonia. "Catorce municipios del Sur de la Amazonia ya cuentan con Determinantes Ambientales definidas y actualizadas." *Corpoamazonia*, May 7, 2019. www.corpoamazonia.gov.co/index.php/2-principal/1026-catorce-municipios-amazonia-cuentan-con-determinantes-ambientales?Fbclid=IwAR1rOBT-Oanra706pnb6SHFBHFUDqlp82Q5Abj1GX3ViTyH6CG1OzlN1VtI.

Corte Constitutional. "Sentencia T-622/16." November 10, 2016. https://www.corteconstitucional.gov.co/relatoria/2016/t-622-16.htm.

Corte Suprema Justica. "STC4360-2018." April 5, 2018. www.observatoriop10.cepal.org/sites/default/files/documents/stc4360-2018.pdf.

Costanza-Chock, Sasha. *Design Justice: Community-Led Practices to Build the Worlds We Need*. Cambridge, MA: MIT Press, 2020.

Coverdale, Helen B. "Caring and the Prison in Philosophy, Policy, and Practice: Under Lock and Key." *Journal of Applied Philosophy* 38, no. 3 (2021): 415–30.

Crapanzano, Vincent. *Imaginative Horizons*. Chicago: University of Chicago Press, 2010.

Crenshaw, Kimberlé. "Demarginalizing the Intersection of Race and Sex: A Black Feminist Critique of Antidiscrimination Doctrine, Feminist Theory and Antiracist Politics." *University of Chicago Legal Forum* 1, no. 8 (1989): 139–67.

Cripps, Elizabeth. "Saving the Polar Bear, Saving the World: Can the Capabilities Approach Do Justice to Humans, Animals and Ecosystems?" *Res Publica* 16, no. 1 (2010): 1–22.

Cummins, Eric. *The Rise and Fall of California's Radical Prison Movement*. Stanford, CA: Stanford University Press, 1994.

Dacudao, Patricia I. "Abaca: The Socio-Economic and Cultural Transformation of Frontier Davao, 1898–1941." PhD diss., Murdoch University, 2017.

Dave, Naisargi N. "What It Feels Like to Be Free: The Tense of Justice." *Fieldsights: Society for Cultural Anthropology*, June 29, 2018. www.culanth.org/fieldsights/what-it-feels-like-to-be-free-the-tense-of-justice.

David, Randy S., Temario C. Rivera, Patricio Abinales, Oliver G. Teves, Jr. Resabal, and S. Procopio. *Transnational Corporations and the Philippine Banana Export Industry*. Quezon City: University of the Philippines Third World Studies Center, 1981.

Davis, Angela Y. *Are Prisons Obsolete?* New York: Seven Stories Press, 2011.

Davis, Angela Y. *Freedom Is a Constant Struggle: Ferguson, Palestine, and the Foundations of a Movement*. Chicago: Haymarket Books, 2016.

Davis, Heather, and Zoe Todd. "On the Importance of a Date, or Decolonizing the Anthropocene." *ACME: An International Journal for Critical Geographies* 116, no. 4 (2017): 761–80.

Davis, Janae, Alex A. Moulton, Levi Van Sant, and Brian Williams. "Anthropocene, Capitalocene, … Plantationocene?: A Manifesto for Ecological Justice in an Age of Global Crises." *Geography Compass* 13, no. 5 (2019): 1–15.

Davis, Mike. "Utah's Toxic Heaven." *Capitalism Nature Socialism* 9, no. 2 (1998): 35–39.

De, Rohit. "Cows and Constitutionalism." *Modern Asian Studies* 53, no. 1 (2019): 240–77.

Dejusticia. "The Colombian Government Has Failed to Fulfill the Supreme Court's Landmark Order to Protect the Amazon." *Dejusticia*, April 5, 2019. www.dejusticia.org/en/the-colombian-government-has-failed-to-fulfill-the-supreme-courts-landmark-order-to-protect-the-amazon/.

De la Cadena, Marisol. *Earth Beings: Ecologies of Practice across Andean Worlds*. Durham, NC: Duke University Press, 2015.

De la Cadena, Marisol. "An Interview with Marisol de la Cadena." Interview by Yoko Taguchi. *NatureCulture*, December 2016. Accessed October 20, 2020. www.natcult.net/interviews/an-interview-with-marisol-de-la-cadena/.

De la Cadena, Marisol. "An Invitation to Live Together, Making the 'Complex We.'" *Environmental Humanities* 11, no. 2 (2019): 477–84.

De la Cadena, Marisol, and Mario Blaser, eds. *A World of Many Worlds*. Durham, NC: Duke University Press, 2018.

Demos, T. J., Emily Eliza Scott, and Subhankar Banerjee, eds. *The Routledge Companion to Contemporary Art, Visual Culture, and Climate Change*. New York: Routledge, 2021.

De Oliveira Andreotti, Vanessa, Sharon Stein, Cash Ahenakew, and Dallas Hunt. "Mapping Interpretations of Decolonization in the Context of Higher Education." *Decolonization: Indigeneity, Education and Society* 4, no. 1 (2015): 21–40.

Derrida, Jacques. "For a Justice to Come: Interview with Jacques Derrida." By Lieven De Canter. In *The Derrida-Habermas Reader*, edited by Lasse Thomassen, 259–69. Edinburgh: Edinburgh University Press.

Derrida, Jacques. "Force of Law: The 'Mystical Foundation of Authority.'" In *Deconstruction and the Possibility of Justice*, edited by Drucilla Cornell, Michel Rosenfeld, and David G. Carlson, 3–67. New York: Routledge, 1992.

Derrida, Jacques. *The Gift of Death*. Chicago: University of Chicago Press, 2017.

Derrida, Jacques. "Marx and Sons." In *Ghostly Demarcations: A Symposium on Jacques Derrida's Specters of Marx*, edited by Michael Sprinker, 213–69. New York: Verso, 1999.

Derrida, Jacques. *Specters of Marx: The State of the Debt, the Work of Mourning, and the New International*. Translated by Peggy Kamuf. New York: Routledge, 1994.

Derrida, Jacques. "Step of Hospitality/No Hospitality." In *Of Hospitality: Anne Dufourmantelle Invites Jacques Derrida to Respond*, 75–160. Translated by Rachel Bowlby. Stanford, CA: Stanford University Press, 2000.

Dhar, Murli, and Surinder Jodhka. "Cow, Caste and Communal Politics: Dalit Killings in Jhajjar." *Economic and Political Weekly* 38, no. 3 (2003): 174–76.

Dial, Roman, and Jonathan Roughgarden. "Theory of Marine Communities: The Intermediate Disturbance Hypothesis." *Ecology* 79, no. 4 (1998): 1412–24.

Díaz Parra, Karla. *Entre el Estado unitario y la autonomía territorial: Implementación de los principios de coordinación y concurrencia por parte del sector petrolero en el piedemonte amazónico*. Bogotá: Asociación Ambiente y Sociedad, 2021.

Díaz Parra, Karla, and M. A. Aguilar Herrera. *Ordenamiento territorial y ambiental de la Amazonía colombiana en el posconflicto*. Bogotá: Asociación Ambiente y Sociedad, 2018.

Douglas, Leah. "Mapping Covid-19 Outbreaks in the Food System." *Food and Environment Reporting Network*, April 22, 2020. www.thefern.org/2020/04/mapping-covid-19-in-meat-and-food-processing-plants/.

Douglass, Frederick. *Narrative of the Life of Frederick Douglass, an American Slave*. Cambridge, MA: Harvard University Press, 2009.

Duarte, Carlos. *Desencuentros Territoriales: Caracterización de los Conflictos en las Regiones de la Altillanura, Putumayo y Montes de María*. Bogotá: Instituto Colombiano Antropología e Historia, 2016.

Dumit, Joseph. "Writing the Implosion: Teaching the World One Thing at a Time." *Cultural Anthropology* 29, no. 2 (2014): 344–62.

Dupré, John. *The Disorder of Things: Metaphysical Foundations of the Disunity of Science*. Cambridge, MA: Harvard University Press, 1993.

Egelko, Bob. "San Quentin Coronavirus Outbreak Apparently Result of Missed Steps by Prison Overseer." *San Francisco Chronicle*, July 20, 2020. www.sfchronicle.com/news/article/San-Quentin-coronavirus-outbreak-apparently-15421664.Php/.

Eglash, Albert. "Creative Restitution: Its Roots in Psychiatry, Religion and Law." *British Journal of Delinquency* 10 (1959): 114–19.

Eglash, Ron. "An Introduction to Generative Justice." *Teknokultura* 13, no. 2 (2016): 369–404.

Eglash, Ron, and Ellen K. Foster. "On the Politics of Generative Justice: African Traditions and Maker Communities." In *What Do Science, Technology, and Innovation Mean from Africa?*, edited by Clapperton C. Mavhunga, 117–35. Cambridge, MA: MIT Press, 2017.

El Tiempo. "En tres meses, 120 líderes sociales han sido asesinados en Colombia." *El Tiempo*. Accessed May 18, 2020. www.eltiempo.com/colombia/otras-ciudades/el-mapa-de-los-lideres-sociales-asesinados-en-colombia-184408.

Endres, Danielle. "From Wasteland to Waste Site: The Role of Discourse in Nuclear Power's Environmental Injustices." *Local Environment* 14, no. 10 (2009): 917–37.

Endres, Danielle. "Sacred Land or National Sacrifice Zone: The Role of Values in the Yucca Mountain Participation Process." *Environmental Communication* 6, no. 3 (2012): 328–45.

Escobar, Arturo. *Pluriversal Politics: The Real and the Possible*. Durham, NC: Duke University Press, 2020.

ESCR-Net. "STC 4360-2018." April 5, 2018. www.escr-net.org/caselaw/2019/stc-4360-2018.

Estremera, Stella A. "Estremera: I Am Not a Banana." *SunStar Davao*, September 17, 2006.

Estupiñán Achury, Liliana. "Neconstitucionalismo ambiental y derechos de la Naturaleza en el marco del nuevo constitucionalismo Latinoamericano." In *La Naturaleza como sujeto de derechos en el constitucionalismo democrático*, edited by Liliana Estupiñán Achury, Claudia Storini, Rubén Martínez Dalmau, and Fernando Antonio de Carvalho Dantas, 365–87. Bogotá: Universidad Libre, 2019.

Fanon, Frantz. *Black Skin, White Masks*. New York: Grove Press, 1952.

Fanon, Frantz. *The Wretched of the Earth*. New York: Grove Press, 1963.

Farrier, Alan, Michelle Baybutt, and Mark Dooris. "Mental Health and Wellbeing Benefits from a Prison's Horticultural Programme." *International Journal of Prisoner Health* 15, no. 1 (2019): 91–104.

Fernando, Mayanthi. "Supernatureculture." *The Immanent Frame*, December 11, 2017. Accessed September 20, 2020. www.Tif.ssrc.org/2017/12/11/supernatureculture/.

Fikret, Huseynov E., Ahbarov S. Huseyn, and Aliyev S. Mammad. "Planning Sustainable Social Infrastructure in the Green New Cities of Azerbaijan." In *Design for Innovative Value Towards a Sustainable Society*, edited by Mitsutaka Matsumoto, Yasushi Umeda, Keijiro Masui, and Shinichi Fukushige, 1044–48. New York: Springer, 2012.

Fineman, Martha Albertson. *The Autonomy Myth: A Theory of Dependency*. New York: New Press, 2004.

Fineman, Martha Albertson. "Equality, Autonomy, and the Vulnerable Subject in Law and Politics." In *Vulnerability: Reflections on a New Ethical Foundation for Law and Politics*, edited by Martha Albertson Fineman and Anna Grear, 25–40. New York: Routledge, 2016.

Fiol, Stefan. "Dual Framing: Locating Authenticities in the Music Videos of Himalayan Possession Rituals." *Ethnomusicology* 54, no. 1 (2010): 28–53.

Fitz-Henry, Erin. "Distribution without Representation? Beyond the Rights of Na-

ture in the Southern Ecuadorian Highlands." *Journal of Human Rights and the Environment* 12, no. 1 (2021): 5–23.
Fortun, Kim. *Advocacy after Bhopal: Environmentalism, Disaster, New Global Orders.* Chicago: University of Chicago Press, 2001.
Fortun, Kim. "Ethnography in Late Industrialism." *Cultural Anthropology* 27, no. 3 (2012): 446–64.
Foster, John B. "Late Soviet Ecology and the Planetary Crisis." *Monthly Review* 67, no. 2 (2015): 1–20.
Foucault, Michel. *Discipline and Punish: The Birth of the Prison.* London: Allen Lane, 1977.
Foucault, Michel. *A History of Sexuality: An Introduction.* Vol. 1. New York: Vintage, 1985.
Foucault, Michel. "Of Other Spaces: Utopias and Heterotopias." In *Rethinking Architecture: A Reader in Cultural Theory*, edited by Neil Leach, 330–36. London: Routledge, 1997.
Foucault, Michel. *"Society Must Be Defended": Lectures at the Collège de France, 1975–1976.* New York: Picador, 1977.
Freccero, Carla. "Figural Historiography: Dogs, Humans, and Cynanthropic Becomings." In *Comparatively Queer*, edited by Jarrod Hayes, Margaret Higonnet, and William J. Spurlin, 45–67. New York: Palgrave, 2010.
Freitag, Sandria B. "Sacred Symbol as Mobilizing Ideology: The North Indian Search for a 'Hindu' Community." *Comparative Studies in Society and History* 22, no. 4 (1980): 597–625.
Fungicide Resistance Action Committee. "2016 Meeting Minutes–English." 2016. https://www.frac.info/docs/default-source/working-groups/banana-group/group/2016-meeting-minutes---english.pdf?sfvrsn=5cfe4a9a_2.
Gabardi, Wayne. *The Next Social Contract: Animals, the Anthropocene, and Biopolitics.* Philadelphia: Temple University Press, 2017.
Galt, Ryan E. *Food Systems in an Unequal World: Pesticides, Vegetables, and Agrarian Capitalism in Costa Rica.* Tucson: University of Arizona Press, 2014.
Galvin, Shaila Sheshia. "Interspecies Relations and Agrarian Worlds." *Annual Review of Anthropology* 47, no. 1 (2018): 233–49.
Garcia, D., and G. Lovink. "The ABC of Tactical Media." Nettime Mailing List, May 16, 1997. Accessed October 14, 2020. www.nettime.org/Lists-Archives/nettime-l-9705/msg00096.html.
García, María-Elena. *Gastropolitics and the Specters of Race: Stories of Capital, Culture, and Coloniality in Peru.* Berkeley: University of California Press, 2021.
García, María-Elena. "Landscapes of Death: Political Violence beyond the Human in the Peruvian Andes." Unpublished paper. Simpson Center Society of Scholars Workshop, University of Washington.
García Arbeláez, Carolina. "Los jueces del fin del mundo." *Semana Sostenible*, April 26, 2018. www.Sostenibilidad.semana.com/opinion/articulo/carolina-garcia-opinion-semana-sostenible-los-jueces-del-fin-del-mundo/40885.
Garth, Hanna, and Ashanté M. Reese, eds. *Black Food Matters: Racial Justice in the Wake of Food Justice.* Minneapolis: University of Minnesota Press, 2020.

Garzón, Camilo A. "Los Derechos de la Naturaleza se Sintonizan con la Conciencia Ambiental de Nuestro Tiempo." *La Silla Vacia*, February 14, 2020. www.lasillavacia.com/silla-academica/universidad-del-rosario/los-derechos-naturaleza-se-sintonizan-conciencia-ambiental.

Gauthier, David. *Morals by Agreement*. Oxford: Oxford University Press, 1986.

Gay, Ross. *The Book of Delights: Essays*. New York: Algonquin Books, 2019.

Gay, Ross. "Tending Joy and Practicing Delight." *The On Being Project*, July 25, 2019. Last updated March 26, 2020. www.onbeing.org/programs/ross-gay-tending-joy-and-practicing-delight/.

Geertz, Clifford. *Local Knowledge: Further Essays in Interpretive Anthropology*. New York: Basic Books, 1983.

Gell, Alfred. "Vogel's Net: Traps as Artworks and Artworks as Traps." *Journal of Material Culture* 1, no. 1 (1996): 15–38.

Gerber, Michele S. *On the Home Front: The Cold War Legacy of the Hanford Nuclear Site*. 3rd ed. Lincoln: University of Nebraska Press, 2007.

Ghosh, Sahana. "*Chor*, Police and Cattle: The Political Economies of Bovine Value in the India-Bangladesh Borderlands." *South Asia: Journal of South Asian Studies* 42, no. 6 (2019): 1108–24.

Giesen, James C. "'The Herald of Prosperity': Tracing the Boll Weevil Myth in Alabama." *Agricultural History* 85, no. 1 (2011): 24–49.

Gillespie, Katie. *The Cow with Ear Tag #1389*. Chicago: University of Chicago Press, 2018.

Gilligan, Carol. "Remapping the Moral Domain: New Images of Self in Relationship." In *Mapping the Moral Domain: A Contribution of Women's Thinking to Psychological Theory and Education*, edited by Carol Gilligan, Janie Victoria Ward, Jill McLean Taylor, and Betty Bardige, 1–19. Cambridge, MA: Harvard University Press, 1988.

Gilmore, Ruth Wilson. *Abolition Geography: Essays Towards Liberation*. Edited by Brenna Bhandar and Alberto Toscano. New York: Verso, 2022.

Gilmore, Ruth Wilson. "Abolition Geography and the Problem of Innocence." In *Futures of Black Radicalism*, edited by Gaye Theresa Johnson and Alex Lubin, 224–41. New York: Verso, 2017.

Gilmore, Ruth Wilson. *Golden Gulag: Prisons, Surplus, Crisis, and Opposition in Globalizing California*. Berkeley: University of California Press, 2007.

Gilmore, Ruth Wilson. "Literature for Justice: A Path Forward." Presented at the National Book Foundation, December 2, 2020. www.youtu.be/i5YYBwEqCKU.

Gilmore, Ruth Wilson, and Naomi Murakawa. "Covid 19, Decarceration, and Abolition (Full)." Haymarket Books, April 28 2020. Accessed September 20, 2020. www.youtube.com/watch?v=hf3f5i9vJNM.

Giraud, Eva H. *What Comes after Entanglement? Activism, Anthropocentrism, and An Ethics of Exclusion*. Durham, NC: Duke University Press, 2019.

Glick, Megan H. *Infrahumanisms*. Durham, NC: Duke University Press, 2018.

Gold, Ann. "Spirit Possession Perceived and Performed in Rural Rajasthan." *Contributions to Indian Sociology* 22, no. 1 (1988): 35–63.

Gómez-Barris, Macarena. "Decolonial Futures." *Social Text* Periscope, June 7,

2018. Accessed October 20, 2020. www.socialtextjournal.org/periscope_article/decolonial-futures/.

Gómez-Barris, Macarena. *The Extractive Zone: Social Ecologies and Decolonial Perspectives*. Durham, NC: Duke University Press, 2017.

Gómez-Rey, Andrés, Ivan Vargas-Chaves, and Adolfo Ibañez-Elan. "El caso de la Naturaleza: derechos sobre la mesa. ¿Decálogo o herramienta?" In *La Naturaleza como sujeto de derechos en el constitucionalismo democrático*, edited by Liliana Estupiñán Achury, Claudia Storini, Rubén Martínez Dalmau, and Fernando Antonio de Carvalho Dantas, 423–44. Bogotá: Universidad Libre, 2019.

Gonzalez, Carmen. "Bridging the North-South Divide: International Environmental Law in the Anthropocene." *Pace Environmental Law Review* 32 (2015): 407–33.

Gordon, Avery. *Ghostly Matters: Haunting and the Sociological Imagination*. Minneapolis: University of Minnesota Press, 2008.

Gossett, Che. "Blackness, Animality, and the Unsovereign." *Verso* (blog), September 8, 2015. www.versobooks.com/blogs/2228-che-gossett-blackness-animality-and-the-unsovereign.

Gould, Chandr'e, and Peter I. Folb. "Project Coast: Apartheid's Chemical and Biological Warfare Programme." Geneva: United Nations Institute for Disarmament Research (UNIDIR) and Centre for Conflict Resolution (CCR), 2002.

Governor's Office of Indian Affairs. "Treaty of Walla Walla, 1855." Accessed January 26, 2021. https://goia.wa.gov/tribal-government/treaty-walla-walla-1855.

Govindrajan, Radhika. "Adulterous Dotiyal or Protector of the Oppressed? Modernity and the Reframing of Ganganath's *Itihas* in Uttarakhand." In *Religion and Modernity in the Himalayas*, edited by Megan A. Sijapati and Jessica V. Birkenholtz, 107–26. New York: Routledge, 2016.

Govindrajan, Radhika. *Animal Intimacies: Interspecies Relatedness in India's Central Himalayas*. Chicago: University of Chicago Press, 2018.

Govindrajan, Radhika. "Labors of Love: On the Political Economies and Ethics of Bovine Politics in Himalayan India." *Cultural Anthropology* 36, no. 2 (2021): 193–221.

Grant, Paul B. C., Million B. Woudneh, and Peter S. Ross. "Pesticides in Blood from Spectacled Caiman (Caiman Crocodilus) Downstream of Banana Plantations in Costa Rica." *Environmental Toxicology and Chemistry* 32, no. 11 (2013): 2576–83.

Gray, Ros, and Shela Sheikh. "The Wretched Earth: Botanical Conflicts and Artistic Interventions." *Third Text* 32, nos. 2–3 (2018): 163–75.

Greenberg, Jessica R. "When Is Justice Done?" *Fieldsights: Society for Cultural Anthropology*, June 9, 2018. www.culanth.org/fieldsights/when-is-justice-done.

Grossman, Charles M., Rudi H. Nussbaum, and Fred D. Nussbaum. "Thyrotoxicosis among Hanford, Washington, Downwinders: A Community-Based Health Survey." *Archives of Environmental Health* 57, no. 1 (2002): 9–15.

Gumbs, Alexis Pauline. "Freedom Seeds: Growing Abolition in Durham, North Carolina." In *Abolition Now!: Ten Years of Strategy and Struggle against the Prison Industrial Complex*, edited by The CR10 Publications Collective, 145–55. Oakland, CA: AK Press, 2008.

Gumbs, Alexis Pauline. *Spill: Scenes of Black Feminist Fugitivity*. Durham, NC: Duke University Press, 2016.

Gundimeda, Sambaiah. "Democratisation of the Public Sphere: The Beef Stall Case in Hyderabad's *Sukoon* Festival." *South Asia Research* 29, no. 2 (2009): 127–49.

Günel, Gökçe. *Spaceship in the Desert: Energy, Climate Change, and Urban Design in Abu Dhabi*. Durham, NC: Duke University Press, 2019.

Gupta, Charu. "The Icon of Mother in Late Colonial North India." *Economic and Political Weekly* 36, no. 45 (2001): 4291–99.

Gururani, Shubhra. "Forests of Pleasure and Pain: Gendered Practices of Labor and Livelihood in the Forests of the Kumaon Himalayas, India." *Gender, Place, and Culture: A Journal of Feminist Geography* 9, no. 3 (2002): 229–43.

Guthman, Julie. "Lives Versus Livelihoods? Deepening the Regulatory Debates on Soil Fumigants in California's Strawberry Industry." *Antipode* 49, no. 1 (2017): 86–105.

Guthman, Julie, and Sandy Brown. "Whose Life Counts: Biopolitics and the 'Bright Line' of Chloropicrin Mitigation in California's Strawberry Industry." *Science, Technology, and Human Values* 41, no. 3 (2016): 461–82.

Hadidian, John. "Taking the 'Pest' out of Pest Control: Humaneness and Wildlife Damage Management." *Attitudes towards Animals Collection* 14 (2012): 7–11.

Hadidian, John, Bernard Unti, and John Griffin. "Measuring Humaneness: Can It Be Done, and What Does It Mean If It Can?" *Humane Treatment of Animals Collection* 1 (2014): 443–48.

Halberstam, Jack. "The Wild Beyond: With and for the Undercommons." In *The Undercommons: Fugitive Planning and Black Study*, edited by Stefano Harney and Fred Moten, 2–13. Minor Compositions, 2013.

Hall, Stuart. "On Postmodernism and Articulation: An Interview with Stuart Hall." Interview by Lawrence Grossberg. *Journal of Communication Inquiry* 10, no. 2 (1986): 45–60.

Hall, Stuart. "Signification, Representation, Ideology: Althusser and the Post-Structuralist Debates." *Critical Studies in Media Communication* 2, no. 2 (1985): 91–114.

Han, Ah-Reum, Sin-Ae Park, and Byung-Eun Ahn. "Reduced Stress and Improved Physical Functional Ability in Elderly with Mental Health Problems Following a Horticultural Therapy Program." *Complementary Therapies in Medicine* 38 (2018): 19–23.

Hanson, Randel D. "Nuclear Agreement Continues U.S. Policy of Dumping on Goshutes." *The Circle: News from an American Indian Perspective*, January 31, 1995, 8.

Haraway, Donna J. *The Companion Species Manifesto: Dogs, People, and Significant Otherness*. Chicago: Prickly Paradigm Press, 2003.

Haraway, Donna J. *Modest_Witness@Second_Millennium.FemaleMan_Meets_OncoMouse: Feminism and Technoscience*. New York: Routledge, 1997.

Haraway, Donna J. *Simians, Cyborgs, and Women: The Reinvention of Nature*. London: Free Association Books, 1991.

Haraway, Donna J. *Staying with the Trouble: Making Kin in the Chthulucene*. Durham, NC: Duke University Press, 2016.

Haraway, Donna J. *When Species Meet*. Minneapolis: University of Minnesota Press, 2007.

Harding, Sandra. "'Strong Objectivity': A Response to the New Objectivity Question." *Synthese* 104, no. 3 (1995): 331–49.

Harris-Brandts, Suzanne, and David Gogishvili. "Architectural Rumors: Unrealized Megaprojects in Baku, Azerbaijan and Their Politico-Economic Uses." *Eurasian Geography and Economics* 59, no. 1 (2018): 73–97.

Harrison, Jill L. *Pesticide Drift and the Pursuit of Environmental Justice*. Cambridge, MA: MIT Press, 2011.

Harrison, Robert Pogue. *Gardens: An Essay on the Human Condition*. Chicago: University of Chicago Press, 2009.

Hartman, Saidiya. *Lose Your Mother: A Journey along the Atlantic Slave Route*. New York: Farrar, Straus, and Giroux, 2008.

Harvey, David. *History of the Hanford Site: 1943–1990*. Richland, WA: Pacific Northwest National Laboratory, 2000.

Hatch, Anthony Ryan. *Silent Cells: The Secret Drugging of Captive America*. Minneapolis: University of Minnesota Press, 2019.

Hatch, Anthony Ryan. "Two Meditations in Coronatime." *Science, Knowledge, and Technology Section of the American Sociological Association*, May 22, 2020. www.asaskat.com/2020/05/22/two-meditations-in-coronatime/.

Hauʻofa, Epeli. *We Are the Ocean: Selected Works*. Honolulu: University of Hawaiʻi Press, 2008.

Hazelett, Evan. "Greening the Cage: Racial Capitalism, Carceral Logics, and Moments of Resistance in the (Un)Sustainable Prison Garden." PhD diss., Harvard University, 2020.

Heath-Thornton, Debra. "Restorative Justice and Spirituality." *Faculty Scholarship Papers* 36 (2009).

Hegel, Georg W. F. *Elements of the Philosophy of Right*. Edited by Allen W. Wood. Translated by Allan B. Nisbet. Cambridge: Cambridge University Press, 1991.

Hegel, Georg W. F. *The Phenomenology of Spirit*. Translated by Terry Pinkard. Cambridge: Cambridge University Press, 2018.

Heidegger, Martin. *Being and Time*. Translated by John Macquarrie and Edward Robinson. London: SCM Press, 1962.

Heise, Ursula. *Imagining Extinction: The Cultural Meanings of Endangered Species*. Chicago: University of Chicago Press, 2016.

Heise, Ursula. "Multispecies Justice." Centre for the Humanities, Utrecht University, October 1, 2015. www.youtube.com/watch?v=O8V34sfkdz8/.

Hernawan, Budi. "Torture in Papua: A Spectacle of Dialectics of the Abject and the Sovereign." In *Comprehending West Papua*, edited by Peter King, Jim Elmslie, and Camellia Webb-Gannon, 339–57. Sydney: Centre for Peace and Conflict Studies, University of Sydney, 2011.

Hetherington, Kregg. *Infrastructure, Environment, and Life in the Anthropocene*. Durham, NC: Duke University Press, 2018.

Heynen, Nik, and Megan Ybarra. "On Abolition Ecologies and Making 'Freedom as a Place.'" *Antipode* 53, no. 1 (2021): 21–35.

Hooks, Gregory, and Wendy Sawyer. "Mass Incarceration, COVID-19, and Community Spread." Prison Policy Initiative, December 2020. www.prisonpolicy.org/reports/covidspread.html.

Howard, Amber-Rose, Brian Kaneda, Felicia Gomez, Liz Blum, Julie Mello, Melissa Rowlett, Fatimeh Khan, Elizabeth Fraser, and Kelan Thomas. "The People's Plan for Prison Closure." Californians United for a Responsible Budget (CURB), 2021.

Hunt, Ashley. "Art, Abolition, and the University: Ashley Hunt and the Underground Scholars." Presented at Visualizing Abolition, April 5, 2021. www.youtu.be/WxoJBpodSkU/.

Hurlbut, J. Benjamin. "Technologies of Imagination: Secularism, Transhumanism and the Idiom of Progress." In *Religion and Innovation: Antagonists or Partners?*, edited by Donald. A. Yerxa, 213–28. London: Bloomsbury, 2016.

Ibrahimov, Kamil. "Craftsmen's Quarter in Ancient Baku (Rabad) — Analogy of Black City (Outer City)." *IRS Heritage*, no. 8 (2012): 4–11.

ICL Research Team. *Human Cost of Bananas*. Manila: JCDB, 1979.

Imran, Mohammad. "Impact of 'Cow Politics' on Muslim Community: A Case Study of Ghosi Community of North India." *Prabuddha: Journal of Social Equality* 2, no. 1 (2018): 59–74.

Indigenous Action. "Rethinking the Apocalypse: An Indigenous Anti-Futurist Manifesto." March 19, 2020. www.indigenousaction.org/rethinking-the-apocalypse-an-indigenous-anti-futurist-manifesto/.

Infoamazonia. "Indígenas dicen que la sentencia que otorga derechos a la Amazonia los deja por fuera." *El Espectador*, December 2, 2019. www.elespectador.com/noticias/medio-ambiente/indigenas-dicen-que-la-sentencia-que-otorga-derechos-a-la-amazonia-los-deja-por-fuera/.

Infoamazonia. "Se disparó la deforestación en la Amazonia colombiana (otra vez)." *El Espectador*, April 29, 2020. www.elespectador.com/noticias/medio-ambiente/se-disparo-la-deforestacion-en-la-amazonia-colombiana-otra-vez-articulo-917069/.

Ingold, Tim. "Anthropology beyond Humanity." *Journal of the Finnish Anthropological Society* 38, no. 3 (2013): 5–23.

International Programme on Chemical Safety. "Warfarin: Health and Safety Guide." Health and Safety Guide No. 96. Geneva: World Health Organization, 1995.

Irani, Lilly. *Chasing Innovation: Making Entrepreneurial Citizens in Modern India*. Princeton, NJ: Princeton University Press, 2019.

Ishiyama, Noriko. "Environmental Justice and American Indian Sovereignty: Case Study of a Land-Use Conflict in Skull Valley, Utah." *Antipode* 35, no. 1 (2003): 119–39.

Ishiyama, Noriko. *"Giseikuiki" no America: Kakukaihatsu to Senjuminzoku* [America as "Sacrifice Zones": Nuclear Development and Indigenous Peoples]. Tokyo: Iwanami Shoten, 2020.

Ives, Sarah. "'More-Than-Human' and 'Less-Than-Human': Race, Botany, and the Challenge of Multispecies Ethnography." *Catalyst: Feminism, Theory, Technoscience* 5, no. 2 (2019): 1–5.

Izquierdo, Belkis, and Lieselotte Viaene. "Decolonizing Transitional Justice from Indigenous Territories." *Justicia en las Américas*, June 27, 2018. www.dplfblog.com/2018/06/27/decolonizing-transitional-justice-from-indigenous-territories.

Izquierdo, Germán. "Justicia negra: sobrevivientes de un mundo que no se volverá a ver." *Justicia Rural*, September 10, 2019. Accessed October 20, 2020. https://semanarural.com/web/articulo/conozca-como-las-comunidades-afrocolombianas-impartian-justicia-en-el-pacifico-colombiano/1128.

Jackson, George. *Soledad Brother: The Prison Letters of George Jackson*. Chicago: Lawrence Hill Books, 1994.

Jackson, Zakiyyah Iman. *Becoming Human: Matter and Meaning in an Antiblack World*. Sexual Cultures 53. New York: NYU Press, 2020.

Jaffrelot, Christophe. "Hindu Nationalism and the 'Saffronisation of the Public Sphere': An Interview with Christophe Jaffrelot." By Edward Anderson. *Contemporary South Asia* 26, no. 4 (2018): 468–82.

Jaffrelot, Christophe. *The Hindu Nationalist Movement in India*. New York: Columbia University Press, 1996.

Jassal, Aftab. "Divine Politicking: A Rhetorical Approach to Deity Possession in the Himalayas." *Religions* 7, no. 9 (2016): 1–18.

Jiler, James. *Doing Time in the Garden: Life Lessons through Prison Horticulture*. Oakland, CA: New Village Press, 2006.

Jiménez, Alberto Corsín, and Chloe Nahum-Claudel. "The Anthropology of Traps: Concrete Technologies and Theoretical Interfaces." *Journal of Material Culture* 24, no. 4 (2019): 383–400.

Josephson, Paul, Nicolai Dronin, Ruben Mnatsakanian, Aleh Cherp, Dmitry Efremenko, and Vladislav Larin. *An Environmental History of Russia*. Cambridge: Cambridge University Press, 2013.

Jurisdiccion Especial Para la Paz (JEP). "Unidad de Investigación y Acusación de la JEP, 'reconoce como victima silenciosa el medio ambiente.'" Comunicado 009, June 5, 2019. www.jep.gov.co/SiteAssets/Paginas/UIA/sala-de-prensa/Comunicado%20UIA%20-%20009.pdf.

Kafer, Alison. *Feminist, Queer, Crip*. Bloomington: Indiana University Press, 2013.

Kant, Immanuel. "Perpetual Peace: A Philosophical Sketch." In *Political Writings*, edited by Hans S. Reiss, 93–130. Cambridge: Cambridge University Press, 1991.

Kapoor, Shivani. "'Your Mother, You Bury Her': Caste, Carcass, and Politics in Contemporary India." *Pakistan Journal of Historical Studies* 3, no. 1 (2018): 5–30.

Karma, Filep. *Seakan Kitorang Setengah Binatang: Rasialisme Indonesia di Tanah Papua*. Jayapura: Penerbit Deiyai, 2014.

Kauffman, Craig M., and Pamela L. Martin. "Can Rights of Nature Make Development More Sustainable? Why Some Ecuadorian Lawsuits Succeed and Others Fail." *World Development* 92 (2016): 130–42.

Kaye, Thomas N., Kelli Bush, Chad Naugle, and Carri J. LeRoy. "Conservation Projects in Prison: The Case for Engaging Incarcerated Populations in Conservation and Science." *Natural Areas Journal* 35, no. 1 (2015): 90–97.

Kazimova, Arifa. "Azerbaijan's Extravagant Olive Trees." *RadioFreeEurope/RadioLiberty*, March 1, 2020. www.rferl.org/a/Azerbaijans_Extravagant_Olive_Trees/1971187.html.

Keck, Frédéric. *Avian Reservoirs: Virus Hunters and Birdwatchers in Chinese Sentinel Posts*. Durham, NC: Duke University Press, 2020.

Keck, Frédéric. "Livestock Revolution and Ghostly Apparitions: South China as a Sentinel Territory for Influenza Pandemics." *Current Anthropology* 60, August (2019): 251–59.

Keck, Frédéric. "Sentinels for the Environment Birdwatchers in Taiwan and Hong Kong." *China Perspectives* 102, no. 2 (2015): 43–52.

Keck, Frédéric, and Andrew Lakoff. "Sentinel Devices." *Limn* 3 (2013). www.limn.it/articles/preface-sentinel-devices-2/.

Kelley, Robin D. G. *Freedom Dreams: The Black Radical Imagination*. Boston: Beacon Press, 2002.

Kelley, Robin D. G. "What Did Cedric Robinson Mean by Racial Capitalism?" *Boston Review*, January 12, 2017. www.bostonreview.net/race/robin-d-g-kelley-what-did-cedric-robinson-mean-racial-capitalism.

Kim, Alice, Erica Meiners, and Jill Petty, eds. *The Long Term: Resisting Life Sentences Working Toward Freedom*. Chicago: Haymarket Books, 2018.

Kim, Claire Jean. *Dangerous Crossings: Race, Species, and Nature in a Multicultural Age*. Cambridge: Cambridge University Press, 2015.

Kim, Eleana. "Invasive Others and Significant Others: Strange Kinship and Interspecies Ethics Near the Korean Demilitarized Zone." *Social Research: An International Quarterly* 84, no. 1 (2017): 203–20.

Kimmerer, Robin W. *Braiding Sweetgrass*. Minneapolis: Milkweed Editions, 2014.

Kirksey, Eben. "Chemosociality in Multispecies Worlds: Endangered Frogs and Toxic Possibilities in Sydney." *Environmental Humanities* 12, no. 1 (2020): 23–50.

Kirksey, Eben. *Emergent Ecologies*. Durham, NC: Duke University Press, 2015.

Kirksey, Eben. *Freedom in Entangled Worlds: West Papua and the Architecture of Global Power*. Durham, NC: Duke University Press, 2012.

Kirksey, Eben. "Species: A Praxiographic Study." *Journal of the Royal Anthropological Institute* 21, no. 4 (2015): 758–80.

Kirksey, Eben, and Stefan Helmreich. "The Emergence of Multispecies Ethnography." *Cultural Anthropology* 25, no. 4 (2010): 545–76.

Kirksey, Eben, Craig Schuetze, and Stefan Helmreich. "Introduction: Tactics of Multispecies Ethnography." In *The Multispecies Salon*, edited by Eben Kirksey, 1–24. Durham, NC: Duke University Press, 2014.

Kirksey, Eben, Nicholas Shapiro, and Maria Brodine. "Hope in Blasted Landscapes." In *The Multispecies Salon*, edited by Eben Kirksey, 25–63. Durham, NC: Duke University Press, 2014.

Klenk, Rebecca. "Seeing Ghosts." *Ethnography* 22, no. 2 (2004): 229–47.

Klima, Alan. *Ethnography #9*. Durham, NC: Duke University Press, 2019.

Kohn, Eduardo. "How Dogs Dream: Amazonian Natures and the Politics of Transspecies Engagement." *American Ethnologist* 34, no. 1 (2007): 3–24.

Ku, Kuang-Yi. "Kuang-Yi Ku." Accessed October 14, 2020. www.kukuangyi.com/.

Kuletz, Valerie L. *The Tainted Desert: Environmental and Social Ruin in the American West*. New York: Routledge, 1998.

Kumar, Udaya. "The Strange Homeliness of the Night: Spectral Speech and the Dalit Present in C. Ayappan's Stories." *Studies in Humanities and Social Sciences* 17. nos. 1 and 2 (2015): 178–89.

Laclau, Ernesto, and Chantal Mouffe. *Hegemony and Socialist Strategy: Towards a Radical Democratic Politics*. New York: Verso Trade, 2014.

Laduke, Winona. *All Our Relations: Native Struggles for Land and Life*. Cambridge, MA: South End Press, 1999.

Landeen, Dan, and Allen Pinkham. *Salmon and His People: Fish and Fishing in Nez Perce Culture*. Lewiston, ID: Confluence Press, 1999.

Langwick, Stacey A. "A Politics of Habitability: Plants, Healing, and Sovereignty in a Toxic World." *Cultural Anthropology* 33, no. 3 (2018): 415–43.

"La paz nos va a permitir sacar más petróleo de zonas vedadas por el conflicto." *El Espectador*, April 15, 2016. www.elespectador.com/economia/la-paz-nos-va-a-permitir-sacar-mas-petroleo-de-zonas-vedadas-por-el-conflicto-article-627058/.

Latimer, Joanna. "Being Alongside: Rethinking Relations amongst Different Kinds." *Theory, Culture and Society* 30, nos. 7–8 (2013): 77–104.

Latour, Bruno. *An Inquiry into Modes of Existence*. Cambridge, MA: Harvard University Press, 2013.

Latour, Bruno. "On Interobjectivity." *Mind, Culture, and Activity* 3, no. 4 (1996): 228–45.

Latour, Bruno. *War of the Worlds: What About Peace?* Chicago: Prickly Paradigm Press, 2002.

Lear, Jonathan. *Radical Hope: Ethics in the Face of Cultural Devastation*. Cambridge, MA: Harvard University Press, 2006.

Leavitt, John. "Meaning and Feeling in the Anthropology of Emotions." *American Ethnologist* 23, no. 3 (1996): 514–49.

Legislative Analyst's Office. "The 2021–22 Budget: State Correctional Population Outlook." November 19, 2020. Accessed June 1, 2021. www.lao.ca.gov/Publications/Report/4304/.

Li, Tania M. "To Make Live or Let Die? Rural Dispossession and the Protection of Surplus Populations." In "The Point Is to Change It," special issue, *Antipode* 41 (2010): 66–93.

Lincoln, Sarah L. "Notes from Underground: Fugitive Ecology and the Ethics of Place." *Social Dynamics* 44, no. 1 (2018): 128–45.

Lindemuth, Amy L. "Designing Therapeutic Environments for Inmates and Prison Staff in the United States: Precedents and Contemporary Applications." *Journal of Mediterranean Ecology* 8 (2007): 87–97.

Lips, Julius. *The Origin of Things*. New York: A. A. Wyn, Inc., 1947.

Littin, K. E., D. J. Mellor, B. Warburton, and C. T. Eason. "Animal Welfare and Ethical Issues Relevant to the Humane Control of Vertebrate Pests." *New Zealand Veterinary Journal* 52, no. 1 (2011): 1–10.

Livingston, Julia. *Self-Devouring Growth: A Planetary Parable as Told from Southern Africa*. Durham, NC: Duke University Press, 2019.

Locke, John. *Second Treatise of Government*. Bloomington, IN: Hackett, 2016.

Lorimer, Jamie. "Nonhuman Charisma." *Environment and Planning D: Society and Space* 25, no. 5 (2007): 911–32.

Low, Nicholas, and Brendan Gleeson. *Justice, Society and Nature: An Exploration of Political Ecology*. London: Routledge, 2002.

Lowe, Celia. "Viral Clouds: Becoming H5N1 in Indonesia." *Cultural Anthropology* 25, no. 4 (2010): 625–49.

Lunstrum, Elizabeth, Neel Ahuja, Bruce Braun, Rosemary Collard, Patricia J. Lopez, and Rebecca W. Y. Wong. "More-Than-Human and Deeply Human Perspectives on COVID-19." *Antipode* 53, no. 5 (2021): 1503–25.

Luther, Martin. *Briefwechsel, Vol IX: Briefe vom Mai 1531 bis Januar 1534*. Frankfurt: Schriften-Niederlage des Evangelisches Vereins, 1903.

Luther, Martin. *Luther: Letters of Spiritual Counsel*. Translated by Theodore Tappert. Vancouver: Regent College Publishing, 2003.

Luther, Martin. *The Table Talk of Martin Luther*. Translated by William Hazlitt. London: H. G. Bohn, 1857.

Lyons, Kristina. "¿Cómo sería la construcción de una paz territorial?: Iniciativas de justicia socioecológica en el sur." *A la Orilla del Río,* May 25, 2017. www.alaorilladelrio.com/2017/05/25/como-seria-la-construccion-de-una-paz-territorial-iniciativas-de-justicia-socioecologica-en-el-sur.

Lyons, Kristina. *Vital Decomposition: Soil Practitioners and Life Politics*. Durham, NC: Duke University Press, 2020.

Lyons, Nona P. "Two Perspectives: On Self, Relationships, and Morality." *Harvard Educational Review* 53, no. 2 (1983): 125–45.

Macfarlane, Robert. "Should This Tree Have the Same Rights as You?" *Guardian*, November 2, 2019. www.theguardian.com/books/2019/nov/02/trees-have-rights-too-robert-macfarlane-on-the-new-laws-of-nature/.

Macpherson, Elizabeth, and F. Clavijo Ospina. "The Pluralism of River Rights in Aotearoa, New Zealand and Colombia." *Journal of Water Law* 25, no. 6 (2015): 283–93.

Macpherson, Elizabeth, Julia Torres Ventura, and Felipe Clavijo Ospina. "Constitutional Law, Ecosystems, and Indigenous Peoples in Colombia: Biocultural Rights and Legal Subjects." *Transnational Environmental Law* 9, no. 3 (2020): 521–40.

Madley, Benjamin. "California's First Mass Incarceration System: Franciscan Missions, California Indians, and Penal Servitude, 1769–1836." *Pacific Historical Review* 88, no. 1 (2019): 14–47.

Malik, Aditya. "Darbar of Goludev: Possession, Petitions, and Modernity." In *The Law of Possession: Ritual, Healing, and the Secular State*, edited by William S. Sax and Helene Basu, 193–225. New York: Oxford University Press, 2016.

Manapat, Ricardo. *Some Are Smarter than Others: The History of Marcos' Crony Capitalism*. Davao: Aletheia Publications, 1991.

Marder, Michael. "Being Dumped." *Environmental Humanities* 11, no. 1 (2019): 180–93.

Marder, Michael. *Energy Dreams: Of Actuality*. New York: Columbia University Press, 2017.

Marder, Michael. *Hegel's Energy: A Reading of The Phenomenology of Spirit*. Evanston, IL: Northwestern University Press, 2021.

Marder, Michael. "We Couldn't Care More!" In *To Mind Is to Care*, edited by Joke Brouwer and Sjoerd van Tuine, 158–73. Rotterdam: V2 Publishing, 2019.

Marquardt, Steve. "'Green Havoc': Panama Disease, Environmental Change, and Labor Process in the Central American Banana Industry." *American Historical Review* 106, no. 1 (2001): 49–80.

Martineau, Jarrett, and Eric Ritskes. "Fugitive Indigeneity: Reclaiming the Terrain of Decolonial Struggle through Indigenous Art." *Decolonization: Indigeneity, Education and Society* 3, no. 1 (2014): i–xii.

Martinez-Alier, Joan. "The Environmentalism of the Poor: Its Origins and Spread." In *A Companion to Global Environmental History*, edited by J. R. McNeill and Erin Steward Mauldin, 513–29. Hoboken, NJ: Wiley, 2012.

Marty, Martin E. *Martin Luther: A Life*. New York: Penguin, 2008.

Mathews, F. "Bioproportionality: A Necessary Norm for Conservation?" *Journal of Population and Sustainability* 4, no. 1 (2019): 43–53.

Matsuda, Mari J., Charles R. Lawrence III, Richard Delgado, and Kimberlé Williams Crenshaw. *Words That Wound: Critical Race Theory, Assaultive Speech, and the First Amendment*. New York: Routledge, 2018.

Mavhunga, Clapperton C. *The Mobile Workshop: The Tsetse Fly and African Knowledge Production*. Cambridge, MA: MIT Press, 2018.

Mavhunga, Clapperton, C. "Vermin Beings: On Pestiferous Animals and Human Game." *Social Text* 29, no. 1 (2011): 151–76.

Mavhunga, Clapperton C., ed. *What Do Science, Technology, and Innovation Mean from Africa?* Cambridge, MA: MIT Press, 2017.

Mbembe, Achille. "Necropolitics." *Public Culture* 15, no. 1 (2003): 11–40.

Mbembe, Achille. *Necropolitics*. Durham, NC: Duke University Press, 2019.

Mbembe, Achille. *On the Postcolony*. Berkeley: University of California Press, 2001.

Mbembe, Achille, and Carolyn Shread. "The Universal Right to Breathe." *Critical Inquiry* 47, no. S2 (2020): S58–62.

McDonald, CeCe. Foreword to *Captive Genders: Trans Embodiment and the Prison Industrial Complex*, edited by Eric Stanley and Nat Smith, 1–6. 2nd ed. Oakland, CA: AK Press, 2015.

McGaughey, Ewan. *A Casebook on Labour Law*. New York: Hart, 2018.

McKittrick, Katherine. "Plantation Futures." *Small Axe* 17, no. 3 (2013): 1–15.

Mendez, Annelle, Luisa E. Castillo, Clemens Ruepert, Konrad Hungerbuehler, and Carla A. Ng. "Tracking Pesticide Fate in Conventional Banana Cultivation in Costa Rica: A Disconnect between Protecting Ecosystems and Consumer Health." *Science of the Total Environment* 613–14 (2018): 1250–62.

Mercado, Sol. "Not Just a Gardening Program: A Dynamic Approach to Healing, Transformation and Reentry." Paper presented at the Social and Ecological Infrastructure for Recidivism Reduction, March 24, 2021.

Mignolo, Walter, and Catherine Walsh. *On Decoloniality: Concepts, Analytics, Praxis*. Durham, NC: Duke University Press, 2018.

Mill, John Stuart. *Utilitarianism and On Liberty: Including Mill's 'Essay On Bentham' and Selections from the Writings of Jeremy Bentham and John Austin*. Malden, MA: Blackwell, 2008.

Miller, David. *Social Justice*. Oxford: Oxford University Press, 1979.

Millican, Juliet, Carrie Perkins, and Andrew Adam-Bradford. "Gardening in Displacement: The Benefits of Cultivating in Crisis." *Journal of Refugee Studies* 32, no. 3 (2019): 351–71.

Minton, Carol A., Pamela R. Perez, and Kevin Miller. "Voices from behind Prison

Walls: The Impact of Training Service Dogs on Women in Prison." *Society and Animals* 23, no. 5 (2015): 484–501.

Mirzoeff, Nicholas. *The Right to Look*. Durham, NC: Duke University Press, 2011.

Moore, Donald S., Anand Pandian, and Jake Kosek. "The Cultural Politics of Race and Nature: Terrains of Power and Practice." In *Race, Nature, and the Politics of Difference*, edited by Donald Moore, Jake Kosek, and Anand Pandian, 1–70. Durham, NC: Duke University Press, 2003.

Moran, Dominique. "Budgie Smuggling or Doing Bird? Human-Animal Interactions in Carceral Space: Prison(er) Animals as Abject and Subject." *Social and Cultural Geography* 16, no. 6 (2015): 634–53.

Moran, Dominique, Jennifer Turner, and Anna K. Schliehe. "Conceptualizing the Carceral in Carceral Geography." *Progress in Human Geography* 42, no. 5 (2017): 666–86.

Morgan, Jennifer L. "Accounting for the 'Most Excruciating Torment.'" In *Reckoning with Slavery*, 141–69. Durham, NC: Duke University Press, 2021.

Morgan, Lewis Henry. *The American Beaver and His Works*. Philadelphia: J. B. Lippincott and Co., 1868.

Morrill, Angie, Eve Tuck, and the Super Futures Haunt Qollective. "Before Dispossession, or Surviving It." *Liminalities: A Journal of Performance Studies* 12, no. 1 (2016): 1–20.

Morris, Ruth. *Stories of Transformative Justice*. Toronto: Canadian Scholars' Press, 2000.

Moten, Fred. *Stolen Life*. Durham, NC: Duke University Press, 2018.

Muiu, Mueni Wa, and Guy Martin. *A New Paradigm of the African State: Fundi Wa Afrika*. New York: Palgrave Macmillan, 2009.

Muñoz, José Esteban. "Theorizing Queer Inhumanisms: The Sense of Brownness." GLQ: *A Journal of Lesbian and Gay Studies* 21, no. 2 (2015): 209–10.

Mura, Marika Noemi. "The Discontented Farmer: State-Society Relations and Food Insecurity in Rural Tanzania." PhD diss., University of Warwick, 2015.

Murakawa, Naomi. *The First Civil Right: How Liberals Built Prison America*. New York: Oxford University Press, 2014.

Murphy, Michelle. "Afterlife and Decolonial Chemical Relations." *Cultural Anthropology* 32, no. 4 (2017): 494–503.

Murphy, Michelle. "Chemical Regimes of Living." *Environmental History* 13, no. 4 (2008): 695–703.

Mutongi, Kenda. *Matatu: A History of Popular Transportation in Nairobi*. Chicago: University of Chicago Press, 2017.

Myers, Natasha. "From Edenic Apocalypse to Gardens against Eden." In *Infrastructure, Environment, and Life in the Anthropocene*, edited by Kregg Hetherington, 115–48. Durham, NC: Duke University Press, 2015.

Myers, Natasha. "From the Anthropocene to the Planthroposcene: Designing Gardens for Plant/People Involution." *History and Anthropology* 28, no. 3 (2017): 1–5.

Myers, Natasha. *Rendering Life Molecular: Models, Modelers, and Excitable Matter*. Durham, NC: Duke University Press, 2015.

Nading, Alex M., and Josh Fisher. "*Zopilotes, Alacranes, y Hormigas* (Vultures, Scorpions, and Ants): Animal Metaphors as Organizational Politics in a Nicaraguan Garbage Crisis." *Antipode* 50, no. 4 (2017): 997–1015.

Narayanan, Yamini. "Cow Is a Mother, Mothers Can Do Anything for Their Children!: Gaushalas as Landscapes of Anthropatriarchy and Hindu Patriarchy." *Hypatia* 34, no. 2 (2019): 195–221.

National Bureau of Statistics, United Republic of Tanzania. "Gross Domestic Product 2017." March 2018. www.nbs.go.tz/nbs/takwimu/na/GDP_Jan_Dec_2017.pdf.

National Research Council. *Animals as Sentinels of Environmental Health Hazards*. Washington, DC: National Academies Press, 1991.

New York Times. "Covid in the U.S.: Latest Map and Case Count." Accessed July 20, 2020. www.nytimes.com/interactive/2020/us/coronavirus-us-cases.html.

Nikol, Lisette J., and Kees Jansen. *The Philippine Controversy over Aerial Spraying of Pesticides: A Timeline of Selected Developments, 1997–2016*. Wageningen University, Rural Sociology Group. Accessed September 20, 2020. doi:10.18174/442444.

Nikol, Lisette J., and Kees Jansen. "The Politics of Counter-Expertise on Aerial Spraying: Social Movements Denouncing Pesticide Risk Governance in the Philippines." *Journal of Contemporary Asia* 50, no. 1 (2020): 99–124.

Nixon, Rob. *Slow Violence and the Environmentalism of the Poor*. Cambridge, MA: Harvard University Press, 2011.

Nonato, Vince. "SC Voids Ordinance vs Aerial Spraying." *Philippine Daily Inquirer*, August 22, 2016. www.newsinfo.inquirer.net/808262/sc-voids-ordinance-vs-aerial-spraying.

Nozick, Robert. *Anarchy, State and Utopia*. Oxford: Blackwell, 1975.

Nussbaum, Martha Craven. "Creating Capabilities: The Human Development Approach and Its Implementation." *Hypatia* 24, no. 3 (2009): 211–15.

Nussbaum, Martha Craven. *Frontiers of Justice: Disability, Nationality, Species Membership*. Cambridge, MA: Belknap Press, 2006.

Nussbaum, Martha, and Amartya Sen, eds. *The Quality of Life: WIDER Studies in Development Economics*. New York: Oxford University Press, 1993.

O'Brien, Mary. *Making Better Environmental Decisions: An Alternative to Risk Assessment*. Cambridge, MA: MIT Press, 2000.

Ogden, Laura A., Billy Hall, and Kimiko Tanita. "Animals, Plants, People, and Things: A Review of Multispecies Ethnography." *Environment and Society* 4, no. 1 (2013): 5–24.

Okorafor, Nnedi. "Africanfuturism Defined." *Nnedi's Wahala Zone Blog*, October 19, 2019. www.nnedi.blogspot.com/2019/10/africanfuturism-defined.html.

Ong, Aihwa. *Spirits of Resistance and Capitalist Discipline: Factory Women in Malaysia*. Albany: State University of New York Press, 1987.

Oppermann, Serpil. "The Scale of the Anthropocene: Material Ecocritical Reflections." *Mosaic: A Journal for the Interdisciplinary Study of Literature* 51, no. 3 (2018): 1–17.

Ottinger, Gwen. "Changing Knowledge, Local Knowledge, and Knowledge Gaps:

STS Insights into Procedural Justice." *Science, Technology, and Human Values* 38, no. 2 (213): 250–70.

Paredes, Alyssa. "Chemical Cocktails Defy Pathogens and Regulatory Paradigms." In *FeralAtlas: The More-than-Human Anthropocene*, edited by Anna L. Tsing, Jennifer Deger, Alder Keleman, and Feifei Zhou. Stanford, CA: Stanford University Press Digital Projects, 2020. https://feralatlas.supdigital.org/poster/chemical-cocktails-defy-pathogens-and-regulatory-paradigms.

Parreñas, Juno S. *Decolonizing Extinction: The Work of Care in Orangutan Rehabilitation*. Durham, NC: Duke University Press, 2018.

Patterson, Charles. *Eternal Treblinka: Our Treatment of Animals and the Holocaust*. New York: Lantern Books, 2002.

Peet, Richard, and Michael Watts. *Liberation Ecologies: Environment, Development and Social Movements*. New York: Routledge, 2004.

Pellow, David N. *What Is Critical Environmental Justice?* Malden, MA: Polity, 2017.

Platt, Tony. *Beyond These Walls: Rethinking Crime and Punishment in the United States*. New York: St. Martin's Press, 2019.

Plumwood, Val. "Tasteless: Towards a Food-Based Approach to Death." *Environmental Values* 17, no. 3 (2008): 323–30.

Povinelli, Elizabeth. *Geontologies: A Requiem to Late Liberalism*. Durham, NC: Duke University Press, 2016.

Presidencia de la republica. "Con la puesta en marcha de la Campaña 'Artemisa,' buscamos parar la hemorragia deforestadora que se ha visto en los últimos años en el país: Presidente Duque." April 28, 2019. www.id.presidencia.gov.co/Paginas/prensa/2019/190428-puesta-marcha-Campana-Artemisa-buscamos-parar-hemorragia-deforestadora-ha-visto-ultimos-anios-pais-Duque.aspx.

Prison Law Office. "PLO's Efforts to Address COVID-19 in California Prisons." Accessed May 31, 2021. www.prisonlaw.com/cdcr-covid/.

Puar, J. K. *The Right to Maim*. Durham, NC: Duke University Press, 2017.

Puig de la Bellacasa, María. *Matters of Care: Speculative Ethics in More Than Human Worlds*. Minneapolis: University of Minnesota Press, 2017.

Puig de la Bellacasa, María. "'Nothing Comes without Its World': Thinking with Care." *Sociological Review* 60, no. 2 (2012): 197–216.

Raffles, Hugh. *Insectopedia*. New York: Pantheon Books, 2010.

Raffles, Hugh. "Intimate Knowledge." *International Social Science Journal* 54, no. 173 (2002): 325–35.

Raffles, Hugh. "Jews, Lice, and History." *Public Culture* 19, no. 3 (2007): 521–66.

Ralph, Laurence. *The Torture Letters*. Chicago: University of Chicago Press, 2020.

Ramírez, María Clemencia. *Entre el estado y la guerrilla: Identidad y ciudadanía en el movimiento de los campesinos cocaleros del Putumayo*. Bogotá: ICANH, 2001.

Rancière, Jacques. *On the Shores of Politics*. New York: Verso, 1995.

Rapalje, Stewart, and Robert L. Lawrence. *A Dictionary of American and English Law*. 2 vols. Union, NJ: Lawbooks Exchange, 1997.

Rawls, John. *Political Liberalism*. New York: Columbia University Press, 2005.

Rawls, John. *A Theory of Justice: Revised Edition*. Cambridge, MA: Harvard University Press, 1999.

Reardon, Jenny. "On the Emergence of Science and Justice." *Science, Technology, and Human Values* 38, no. 2 (2013): 176–200.

Reardon, Jenny, Jacob Metcalf, Martha Kenney, and Karen Barad. "Science & Justice: The Trouble and The Promise." *Catalyst: Feminism, Theory, Technoscience* 1, no. 1 (2015): 1–48.

Reuben, Suzanne H. *Facing Cancer in Indian Country: The Yakama Nation and Pacific Northwest Tribes, President's Cancer Panel 2002 Annual Report*. Bethesda, MD: U.S. Department of Health and Human Services, 2003. Accessed October 20, 2020. www.deainfo.nci.nih.gov/advisory/pcp/archive/pcp02rpt/YakamaBook.pdf.

Rice, Mark. *Dean Worcester's Fantasy Islands: Photography, Film, and the Colonial Philippines*. Ann Arbor: University of Michigan Press, 2014.

Rice, Stian. "Convicts Are Returning to Farming—Anti-Immigrant Policies Are the Reason." *UMBC Magazine*, June 7, 2019. Accessed October 10, 2020. www.magazine.umbc.edu/convicts-are-returning-to-farming-anti-immigrant-policies-are-the-reason/.

Richie, Beth E. *Arrested Justice: Black Women, Violence, and America's Prison Nation*. New York: New York University Press, 2012.

Richland, Justin B. "Jurisdiction: Grounding Law in Language." *Annual Review of Anthropology* 42 (2013): 209–26.

Robb, Peter. "The Challenge of Gau-Mata: British Policy and Religious Change in India, 1880–1916." *Modern Asian Studies* 20, no. 2 (1986): 285–319.

Robbins, Joel. "Beyond the Suffering Subject: Toward an Anthropology of the Good." *Journal of the Royal Anthropological Institute* 19, no. 3 (2013): 447–62.

Roberts, Dorothy. *Fatal Invention: How Science, Politics, and Big Business Re-Create Race in the Twenty-First Century*. New York: New Press, 2011.

Rodríguez, Gloria Amparo, and Nidia Catherine González. "La jurisdicción especial indígena y los retos del acceso a la justicia ambiental." In *La Naturaleza como sujeto de derechos en el constitucionalismo democrático*, edited by Liliana Estupiñán Achury, Claudia Storini, Rubén Martínez Dalmau, and Fernando Antonio de Carvalho Dantas, 473–94. Bogotá: Universidad Libre, 2019.

Rodríguez Garavito, César, and Diana Rodríguez Franco. *Juicio a la exclusión: El impacto de los tribunales sobre los derechos sociales en el Sur Global*. Buenos Aires: Siglo Veintiuno Editores, 2015.

Rodríguez Garavito, César, Diana Rodríguez Franco, and Helena Durán Crane. *La paz ambiental: retos y propuestas para el posacuerdo*. Bogotá: DeJusticia, 2017.

Romero, Adam M., Julie Guthman, Ryan E. Galt, Matt Huber, Becky Mansfield, and Suzana Sawyer. "Chemical Geographies." *GeoHumanities* 3, no. 1 (2017): 158–77.

Rose, Deborah Bird. "Double Death." *The Multispecies Salon*. Accessed July 28, 2020. www.multispecies-salon.org/double-death/.

Rose, Deborah Bird. *Reports from a Wild Country: Ethics for Decolonisation*. Sydney: University of New South Wales Press, 2004.

Rose, Deborah Bird. *Wild Dog Dreaming: Love and Extinction*. Charlottesville: University of Virginia Press, 2011.

Rose, Deborah Bird, and Thom van Dooren. Introduction to "Unloved Others: Death of the Disregarded in the Time of Extinctions." Special issue. *Australian Humanities Review* 50 (2011): 1–4.

Rose, Nicole. *The Prisoner's Herbal*. London: Active Distribution, 2020.

Rousell, David. "Doing Little Justices: Speculative Propositions for an Immanent Environmental Ethics." *Environmental Education Research* 26, nos. 9–10 (2020): 1391–405.

Rubenstein, Mary-Jane. *Strange Wonder: The Closure of Metaphysics and the Opening of Awe*. New York: Columbia University Press, 2009.

Ruiz Serna, Daniel. "El territorio como víctima: Ontología política y leyes de víctimas para comunidades indígenas y negras en Colombia." *Revista Colombiana de Antropología* 53, no. 2 (2017): 85–113.

Rutherford, Danilyn. *Laughing at Leviathan: Sovereignty and Audience in West Papua*. Chicago: University of Chicago Press, 2012.

Saloner, Brendan, Kalind Parish, Julie Ward, Grace DiLaura, and Sharon Dolovich. "COVID-19 Cases and Deaths in Federal and State Prisons." *Journal of the American Medical Association,* July 8, 2020.

Sax, William. *God of Justice: Ritual Healing and Social Justice in the Central Himalayas*. New York: Oxford University Press, 2009.

Scanlon, Thomas. *What We Owe to Each Other*. Cambridge, MA: Harvard University Press, 2000.

Scarry, Elaine. *On Beauty and Being Just*. Princeton, NJ: Princeton University Press, 2013.

Schaffer, Michael C. *One Nation under Dog: America's Love Affair with Our Dogs*. New York: Henry Holt, 2010.

Scheffler, Samuel. *Equality and Tradition: Questions of Value in Moral and Political Theory*. New York: Oxford University Press, 2010.

Schept, Judah. *Progressive Punishment*. New York: New York University Press, 2015.

Schept, Judah. "(Un)Seeing like a Prison: Counter-Visual Ethnography of the Carceral State." *Theoretical Criminology* 18, no. 2 (2014): 198–223.

Schept, Judah, and Jordan E. Mazurek. "Layers of Violence: Coal Mining, Convict Leasing, and Carceral Tourism in Central Appalachia." In *The Palgrave Handbook of Prison Tourism*, edited by Jacqueline Z. Wilson, Sarah Hodgkinson, Justin Piché, and Kevin Walby, 171–90. Palgrave Studies in Prisons and Penology. London: Palgrave Macmillan UK, 2017.

Schlanger, Margo. "Plata v. Brown and Realignment: Jails, Prisons, Courts, and Politics." *Harvard Civil Rights Civil Liberties Law Review* 48, no. 1 (2013): 165–215.

Schlosberg, David. *Defining Environmental Justice: Theories, Movements, and Nature*. New York: Oxford University Press, 2009.

Schlosberg, David. "Ecological Justice for the Anthropocene." In *Political Animals and Animal Politics*, edited by Marcel Wissenburg and David Schlosberg, 75–89. New York: Palgrave Macmillan, 2014.

Schlosberg, David. "Reconceiving Environmental Justice: Global Movements and Political Theories." *Environmental Politics* 13, no. 3 (2004): 517–40.

Schlosberg, David, and Lisette B. Collins. "From Environmental to Climate Justice:

Climate Change and the Discourse of Environmental Justice." *Wiley Interdisciplinary Reviews: Climate Change* 5, no. 3 (2014): 359–74.

Semana. "Putumayo, clave para el futuro petrolero del país." November 13, 2018. https://www.semana.com/contenidos-editoriales/hidrocarburos-son-el-futuro/articulo/putumayo-clave-para-el-futuro-petrolero-del-pais/590016/.

Sen, Amartya. *The Idea of Justice*. Cambridge, MA: Belknap Press of Harvard University Press, 2011.

Serje, Margarita. *El Revés de la Nación: Territorios Salvajes, Fronteras y Tierras de Nadie*. Bogotá: Universidad de los Andes, 2011.

Serpell, James. *In the Company of Animals: A Study of Human-Animal Relationships*. Cambridge: Cambridge University Press, 1996.

Serres, Michel. *The Parasite*. Minneapolis: University of Minnesota Press, 2007.

Shah, Alpa. "Against Eco-Incarceration : Class Struggle and Indigenous Rights in India by Alpa Shah." *MR Online,* 2013. https://mronline.org/2012/06/07/shah070612-html.

Shange, Savannah. *Progressive Dystopia*. Durham, NC: Duke University Press, 2019.

Shapiro, Nicholas. "Attuning to the Chemosphere: Domestic Formaldehyde, Bodily Reasoning, and the Chemical Sublime." *Cultural Anthropology* 30, no. 3 (2015): 368–93.

Shapiro, Nicholas, and Eben Kirksey. "Chemo-Ethnography: An Introduction." *Cultural Anthropology* 32, no. 4 (2017): 481–93.

Shaviro, Steven. "2015 Conference Plenary Videos." WTF Affect (conference website and documentation). Accessed October 14, 2020. www.wtfaffect.com/steven-shaviro-video/.

Shaviro, Steven. *Discognition*. London: Repeater Books, 2016.

Shepard, P. M. "Advancing Environmental Justice through Community-Based Participatory Research." *Environmental Health Perspectives: Supplements* 110, no. 2 (2002): 139–40. www.ehp.niehs.nih.gov/doi/pdf/10.1289/ehp.02110s2139/.

Shorter, David D. "Spirituality." In *The Oxford Handbook of Indian History*, edited by Frederick E. Hoxie, 433–49. Oxford: Oxford University Press, 2016.

Shrader-Frechette, Kristin. *Environmental Justice: Creating Equality, Reclaiming Democracy*. New York: Oxford University Press, 2002.

Shue, Henry. *Climate Justice: Vulnerability and Protection*. New York: Oxford University Press, 2014.

Shumway, Matthew J., and Richard Jackson. "Place Making, Hazardous Waste, and the Development of Tooele County, Utah." *Geographical Review* 98, no. 4 (2008): 433–55.

Sidgwick, Henry. *The Methods of Ethics*. Cambridge: Cambridge University Press, 2013.

Simpson, Leanne B. "Indigenous Resurgence and Co-Resistance." *Critical Ethnic Studies* 2, no. 2 (2016): 19–34.

Singer, Peter. *Animal Liberation*. New York: Random House, 1990.

Singh, Julietta. *Unthinking Mastery*. Durham, NC: Duke University Press, 2018.

Smeath, Doug. "4 Goshutes Charged with Fraud: Indicted Include Leader of Skull

Valley Band." *Desert News*, December 19, 2003. www.deseret.com/2003/12/19/19802152/4-goshutes-charged-with-fraud/.

Soluri, John. "Accounting for Taste: Export Bananas, Mass Markets, and Panama Disease." *Environmental History* 7, no. 3 (2002): 386–410.

Soluri, John. "People, Plants, and Pathogens: The Eco-Social Dynamics of Export Banana Production in Honduras, 1875–1950." *Hispanic American Historical Review* 80, no. 3 (2000): 463–501.

Sousanis, Nick. *Unflattening*. Cambridge, MA: Harvard University Press, 2015.

Spiegel, Marjorie. *The Dreaded Comparison: Human and Animal Slavery*. [1988]. New York: Mirror Books, 1996.

Spivak, Gayatri C. *Critique of Postcolonial Reason*. Cambridge, MA: Harvard University Press, 1999.

Staples, James. "Blurring Bovine Boundaries: Cow Politics and the Everyday in South India." *South Asia: Journal of South Asian Studies* 42, no. 6 (2019): 1124–40.

Star, Susan Leigh. "Power, Technology and the Phenomenology of Conventions: On Being Allergic to Onions." *Sociological Review* 38, no. 1 (1990): 26–56.

Stengers, Isabelle. "The Cosmopolitical Proposal." In *Making Things Public: Atmospheres of Democracy*, edited by Bruno Latour and Peter Weibel, 994–1003. Cambridge, MA: MIT Press, 2005.

Stengers, Isabelle. *Cosmopolitics I*. Minneapolis: University of Minnesota Press, 2010.

Stengers, Isabelle. *In Catastrophic Times: Resisting the Coming Barbarism*. Open Humanities Press, 2015.

Stengers, Isabelle. "Introductory Notes on an Ecology of Practices." *Cultural Studies Review* 11, no. 1 (2005): 183–96.

Stern, Lesley. "A Garden or a Grave? The Canyonic Landscape of the Tijuana-San Diego Region." In *Arts of Living on a Damaged Planet: Ghosts and Monsters of the Anthropocene*, edited by Anna Lowenhaupt Tsing, Heather Swanson, Elaine Gan, and Nils Bubandt, 17–29. Minneapolis: University of Minnesota Press, 2017.

Stewart-Harawira, Makere. "Returning the Sacred: Indigenous Ontologies in Perilous Times." In *Radical Human Ecology: Intercultural and Indigenous Approaches*, edited by Rose Roberts and Lewis Williams, 94–109. New York: Routledge, 2012.

Stoetzer, Bettina. "Ruderal Ecologies: Rethinking Nature, Migration, and the Urban Landscape in Berlin." *Cultural Anthropology* 33, no. 2 (2018): 295–323.

Story, Brett. *Prison Land: Mapping Carceral Power across Neoliberal America*. Minneapolis: University of Minnesota Press, 2019.

Strange, Stuart. "Vengeful Animals, Involuntary Mourning, and the Ethics of Ndyuka Autonomy." *Cultural Anthropology* 36, no. 1 (2021): 138–65.

Sudbury, Julia, ed. *Global Lockdown: Race, Gender, and the Prison-Industrial Complex*. New York: Routledge, 2004.

Sulek, Julia Prodis, and John Woolfolk. "Coronavirus: San Quentin Prison Guard Fights for Life in San Jose." *Mercury News*, August 1, 2020. www.mercurynews.com/2020/08/01/from-san-quentin-to-san-jose-how-californias-worst-coronavirus-outbreak-burst-from-death-row-to-the-south-bay/.

TallBear, Kim. "Failed Settler Kinship, Truth and Reconciliation, and Science." *Indig-*

enous STS, March 16, 2016. Accessed October 20, 2020. www.indigenoussts.com /failed-settler-kinship-truth-and-reconciliation-and-science/.

TallBear, Kim. "Identity Is a Poor Substitute for Relating: Genetic Ancestry, Critical Polyamory, Property, and Relations." In *Critical Indigenous Studies Handbook*, edited by Brendan Hokowhitu, Linda Tuhiwai-Smith, Chris Andersen, and Steve Larkin, 467–78. New York: Routledge, 2021.

TallBear, Kim. "An Indigenous Reflection on Working beyond the Human/Not Human." *GLQ: A Journal of Lesbian and Gay Studies* 21, nos. 2–3 (2015): 230–35.

TallBear, Kim. "Why Interspecies Thinking Needs Indigenous Standpoints." *Fieldsights: Society for Cultural Anthropology*, November 18, 2011. www.culanth.org /fieldsights/why-interspecies-thinking-needs-indigenous-standpoints/.

Taussig, Michael. *Shamanism, Colonialism and the Wild Man: A Study in Terror and Healing*. Chicago: University of Chicago Press, 1987.

Taylor, Dorceta. *Toxic Communities: Environmental Racism, Industrial Pollution, and Residential Mobility*. New York: New York University Press, 2014.

Taylor, Keeanga-Yamahtta. "The Emerging Movement for Police and Prison Abolition." *New Yorker*, May 7, 2021. www.newyorker.com/news/our-columnists/the -emerging-movement-for-police-and-prison-abolition/.

Tayob, Shaheed. "Disgust as Embodied Critique: Being Middle Class and Muslim in Mumbai." *South Asia: Journal of South Asian Studies* 42, no. 6 (2019): 1192–209.

Teitel, Ruti G. *Transitional Justice*. New York: Oxford University Press, 2000.

Tejano, Ivy C. "Anti-Aerial Spray Group Hopes SC Decision to Change." *SunStar Newspaper*, November 28, 2017. www.sunstar.com.ph/article/407384/.

Temper, Leah, Daniela del Bene, and Joan Martinez-Alier. "Mapping the Frontiers and Front Lines of Global Environmental Justice: The EJAtlas." *Journal of Political Ecology* 22 (2015): 255–78.

Thaler, Mathias. "What If: Multispecies Justice as the Expression of Utopian Desire." *Environmental Politics*. Published ahead of print, March 22, 2021. doi:10.1080 /09644016.2021.1899683.

Thomas, Deborah. "Rights, Gifts, Repair." *American Anthropologist* 121, no. 1 (2020): 5–8.

Thompson, Janna. *Intergenerational Justice: Rights and Responsibilities in an Intergenerational Polity*. New York: Routledge, 2009.

Thompson, Ki'Amber. "Prisons, Policing, and Pollution: Toward an Abolitionist Framework within Environmental Justice." Senior thesis, Pomona College, 2018.

Thuma, Emily L. *All Our Trials: Prisons, Policing, and the Feminist Fight to End Violence*. Urbana: University of Illinois Press, 2019.

Ticktin, Miriam. "Invasive Others: Toward a Contaminated World." *Social Research* 84, no. 1 (2017): xxi–xxxiv.

Ticktin, Miriam. "Invasive Pathogens?: Rethinking Notions of Otherness." *Social Research: An International Quarterly* 84, no. 1 (2017): 55–58.

Todd, Zoe. "Indigenizing the Anthropocene." In *Art in the Anthropocene*, edited by Heather Davis and Etienne Turpin, 241–54. London: Open Humanities Press, 2015.

Tom, Erica. "Humanizing Animals: Talking about Second Chances, Horses, and Prisoners." In *Racial Ecologies*, edited by Leilani Nishime and Kim D. H. Williams, 123–38. Seattle: University of Washington Press, 2018.

Tomassoni, Teresa. "Colombia was the deadliest place on Earth for environmental activists. It's gotten worse." *NBC News*, February 23, 2020. www.nbcnews.com/science/environment/colombia-was-deadliest-place-earth-environmental-activists-it-s-gotten-n1139861.

Tompkins, Kyla W. *Racial Indigestion: Eating Bodies in the 19th Century*. New York: New York University Press, 2012.

Tousignant, Noémi. *Edges of Exposure: Toxicology and the Problem of Capacity in Postcolonial Senegal*. Durham, NC: Duke University Press, 2018.

Transformative In-Prison Workgroup. "TPW." Accessed February 12, 2020. www.thetpw.org/.

Trouillot, Michel-Rolph. "Culture on the Edges: Creolization in the Plantation Context." In *Trouillot Remixed: The Michel-Rolph Trouillot Reader*, edited by Yarimar Bonilla, Greg Beckett, and Mayanthi L. Fernando, 194–214. Durham, NC: Duke University Press, 2021.

Tschakert, Petra, David Schlosberg, Danielle Celermajer, Lauren Rickards, Christine Winter, Mathias Thaler, Makere Stewart-Harawira, and Blanche Verlie. "Multispecies Justice: Climate-Just Futures with, for and beyond Humans." *Wiley Interdisciplinary Reviews: Climate Change* 12, no. 2 (2021): 1–10.

Tsing, Anna L. "Arts of Inclusion, or, How to Love a Mushroom." *Australian Humanities Review*, no. 50 (2011): 5–22.

Tsing, Anna L. "Catachresis for the Anthropocene: Three Papers on Productive Misplacements." *Aura's Openings* 1 (2013): 1–10.

Tsing, Anna L. *The Mushroom at the End of the World: On the Possibility of Life in Capitalist Ruins*. Princeton, NJ: Princeton University Press, 2015.

Tsing, Anna L. "A Threat to Holocene Resurgence Is a Threat to Livability." In *The Anthropology of Sustainability*, edited by Jerome Lewis and Marc Brightman, 51–65. New York: Springer, 2017.

Tsing, Anna L., Jennifer Deger, Alder K. Saxena, and Feifei Zhou. "Feral Atlas: The More-Than-Human Anthropocene." Stanford University Press Digital Projects, 2020.

Tsing, Anna L., Andrew S. Mathews, and Nils Bubandt. "Patchy Anthropocene: Landscape Structure, Multispecies History, and the Retooling of Anthropology. An Introduction to Supplement 20." *Current Anthropology* 60 (2019): 186–97.

Tuck, Eve, and K. Wayne Yang. "What Justice Wants." *Critical Ethnic Studies* 2, no. 2 (2016): 1–15.

Twagira, Laura A. "Introduction: Africanizing the History of Technology." *Technology and Culture* 61, no. 2S (2020): S1–19.

Tylor, Edward Burnett. *Anthropology: An Introduction to the Study of Man and Civilization*. London: Macmillan, 1881.

Uexküll, Jakob von. *A Foray into the Worlds of Animals and Humans, with a Theory of Meaning*. Translated by Joseph D. O'Neil. Minneapolis: University of Minnesota Press, 2010.

United States Environmental Protection Agency. "Superfund: National Priorities

List." Accessed October 7, 2020. www.epa.gov/superfund/superfund-national-priorities-list-npl/.
United States Nuclear Regulatory Commission. "Scoping Meeting for Preparation of an EIS for the Private Fuel Storage Facility." Work Order No.: ASB-300-315. Salt Lake City, UT. June 2, 1998. Accessed January 28, 2021. www.nrc.gov/docs/ML0103/ML010320348.pdf.
Van den Bosch, Robert. *The Pesticide Conspiracy*. Garden City, NY: Anchor, 1980.
van der Schalie, William H., Hank S. Gardner, John A. Bantle, Chris T. De Rosa, Robert A. Finch, John S. Reif, Roy H. Reuter, Lorraine C. Backer, Joanna Burger, Leroy C. Folmar, and William S. Stokes. "Animals as Sentinels of Human Health Hazards of Environmental Chemicals." *Environmental Health Perspectives* 107, no 4 (1999): 309–15.
Van der Veer, Peter. *Religious Nationalism: Hindus and Muslims in India*. Berkeley: University of California Press, 1994.
Van Dooren, Thom, Eben Kirksey, and Ursula Münster. "Multispecies Studies: Cultivating Arts of Attentiveness." *Environmental Humanities* 8, no. 1 (2016): 1–23.
Van Wendel de Joode, Berna, Ana Maria Mora, Leonel Córdoba, Juan Camilo Cano, Rosario Quesada, Moosa Faniband, Catharina Wesseling, Clemens Ruepert, Mattias Oberg, Brenda Eskenazi, Donna Mergler, and Christian H Lindh." "Aerial Application of Mancozeb and Urinary Ethylene Thiourea (ETU) Concentrations among Pregnant Women in Costa Rica: The Infants' Environmental Health Study (ISA)." *Environmental Health Perspectives* 122, no. 12 (2014): 1321–28.
Vasavi, A. R., and Catherine P. Kingfisher. "Poor Women as Economic Agents: The Neo-Liberal State and Gender in India and the US." *Indian Journal of Gender Studies* 10, no. 1 (2003): 1–24.
von Gleich, Paula. "African American Narratives of Captivity and Fugitivity: Developing Post-Slavery Questions for Angela Davis: An Autobiography." *Current Objectives of Postgraduate American Studies* 16, no. 1 (2015): 1–18.
Voskoboynikov, Nikolai. "Mineralogicheskoye Opisaniye Poluostrova Apsherona." *Gornyy zhurnal* 9 (1827): 17–55.
Wagenaar, Henk W., S. S. Parikh, D. F. Plukker, and R. Veldhuijzen van Zanten. *Transliterated Hindi-Hindi-English Dictionary*. New Delhi: Allied Chambers, 1993.
Waitkus, Kathryn E. "The Impact of a Garden Program on the Physical Environment and Social Climate of a Prison Yard at San Quentin State Prison." Master's thesis, Pepperdine University, 2004.
Wang, Jackie. *Carceral Capitalism*. South Pasadena, CA: Semiotext(e), 2018.
Ward, Chip. *Canaries on the Rim: Living Downwind in the West*. New York: Verso, 1999.
Ward, Chip. *The Grantsville Community's Health: A Citizen Survey*. Grantsville, UT: West Desert Healthy Environment Alliance, 1996.
Warren, Dean. *Brazil and the Struggle for Rubber: A Study in Environmental History*. Cambridge: Cambridge University Press, 1987.
Washick, Bonnie, Elizabeth Wingrove, Kathy E. Ferguson, and Jane Bennett. "Politics

That Matter: Thinking about Power and Justice with the New Materialists." *Contemporary Political Theory* 14 (2015): 63–89.

Watkins, Caitlin. "Industrialized Bodies: Women, Food, and Environmental Justice in the Criminal Justice System." In *Addressing Environmental and Food Justice toward Dismantling the School-to-Prison Pipeline: Poisoning and Imprisoning Youth*, edited by Anthony J. Nocella II, John J. Lupinacci, and K. Animashaun Ducre, 137–60. New York: Palgrave Macmillan, 2017.

Watson, Tennessee. "Silent Evidence: One Woman's Story of Childhood Sexual Abuse." *Silent Evidence*, May 19, 2016. www.theheartradio.org/silent-evidence/nottherighttime.

Weaver, Harlan. "Pit Bull Promises: Inhuman Intimacies and Queer Kinships in an Animal Shelter." *GLQ: A Journal of Lesbian and Gay Studies* 21, nos. 2–3 (2015): 343–63.

Weheliye, Alexander G. *Habeas Viscus: Racializing Assemblages, Biopolitics, and Black Feminist Theories of the Human*. Durham, NC: Duke University Press, 2014.

White, Richard. *The Organic Machine: The Remaking of the Columbia River*. New York: Hill and Wang, 1999.

White, Rob, and Hannah Graham. "Greening Justice: Examining the Interfaces of Criminal, Social, and Ecological Justice." *British Journal of Criminology* 55, no. 5 (2015): 845–65.

Whitehead, Alfred North. *Modes of Thought*. New York: Simon and Schuster, 1968.

Whyte, Kyle P. "Indigenous Science (Fiction) for the Anthropocene: Ancestral Dystopias and Fantasies of Climate Change Crises." *Environment and Planning E: Nature and Space* 1, nos. 1–2 (2018): 224–42.

Whyte, Kyle P. "Is It Colonial déjà vu? Indigenous Peoples and Climate Injustice." In *Humanities for the Environment: Integrating Knowledge, Forging New Constellations of Practice*, edited by Joni Adamson and Michael Davis, 88–105. New York: Routledge, 2017.

Whyte, Kyle P. "Our Ancestors' Dystopia Now: Indigenous Conservation and the Anthropocene." In *The Routledge Companion to the Environmental Humanities*, edited by Ursula K. Heise, Jon Christensen, and Michelle Niemann, 206–15. London: Routledge, 2017.

Whyte, Kyle P. "Settler Colonialism, Ecology, and Environmental Injustice." *Environmental and Society: Advances in Research* 9, no. 1 (2018): 125–44.

Wijaya, Elizabeth. "To Learn to Live with Spectral Justice: Derrida-Levinas." *Derrida Today* 5, no. 2 (2012): 232–47.

Wilches Chaux, Gustavo. *Base ambiental para la paz: la necesidad de hacerle gestión de riesgo al paz-conflicto*. Bogotá: Oxfam, 2016.

Wiley, James. *The Banana: Empires, Trade Wars, and Globalization*. Lincoln: University of Nebraska Press, 2008.

Winter, Christine J. "Decolonising Dignity for Inclusive Democracy." *Environmental Values* 28, no. 1 (2019): 9–30.

Wittenberg, Colm. "Poison in the Rhodesian Bush War: How Guerillas Gain Legitimacy." Master's thesis, Leiden University, 2019.

Woodfox, Albert. *Solitary*. New York: Grove Press, 2019.

Woolaston, Katie. "Ecological Vulnerability and the Devolution of Individual Autonomy." *Australasian Journal of Legal Philosophy* 43 (2018): 107–24.

Wynter, Sylvia. "Novel and History, Plot and Plantation." *Savacou* 5 (1971): 95–102.

Wynter, Sylvia. "Unsettling the Coloniality of Being/Power/Truth/Freedom: Towards the Human, after Man, Its Overrepresentation—An Argument." CR: *The New Centennial Review* 3, no. 3 (2003): 257–337.

Yang, Anand A. "Sacred Symbol and Sacred Space in Rural India: Community Mobilization in the 'Anti-Cow Killing' Riot of 1893." *Comparative Studies in Society and History* 22, no. 4 (1980): 576–96.

"Yap Intervenes in Aerial Spray Ban in Davao City." *PhilStar Global*, February 10, 2007. www.philstar.com/nation/2007/02/10/384318/yap-intervenes-aerial-spray-ban-davao-city.

Young, Iris Marion. *Justice and the Politics of Difference*. Princeton, NJ: Princeton University Press, 1990.

Yusoff, Kathryn. *A Billion Black Anthropocenes or None*. Minneapolis: University of Minnesota Press, 2018.

Contributors

Karin Bolender (aka K-Haw Hart) is an interdisciplinary artist and founder of a collaborative, experimental ecological-art platform known as the Rural Alchemy Workshop (R.A.W.). R.A.W. projects seek to cocompose "untold" stories with/in shadowy meshes of mammals, plants, microbes, insects, and elusive others, through expanded practices in performance, writing, video/sound, and book arts. In 2020, 3Ecologies/punctum books published *The Unnaming of Aliass*. As an independent researcher at the seams of contemporary ecological art and environmental humanities, Bolender interweaves experimental creative practices with insights and energies from intersecting scholarly, activist, and other fields of inquiry.

Sophie Chao is a Discovery Early Career Researcher Award (DECRA) Fellow and Lecturer at the Department of Anthropology, University of Sydney. Her anthropological and interdisciplinary research explores the intersections of capitalism, ecology, Indigeneity, health, and justice in the Pacific. Chao previously worked for the international human rights organization Forest Peoples Programme in Indonesia and the United Kingdom. Her book, *In the Shadow of the Palms: More-Than-Human Becomings in West Papua*, was published by Duke University Press in June 2022. For more information, please visit www.morethanhumanworlds.com/.

M. L. Clark is a Canadian speculative-fiction writer now based in Medellín, Colombia. After studying literary histories of science at a doctoral level, Clark left all-but-defense and decided to emigrate. Clark now publishes science fiction with an express focus on social-contract theory and the search for better justices. Her work has been featured in the science fiction magazines *Analog Science Fiction* and *Clarkesworld*, and her first novel is moving through the industry pipeline with *Fantasy&ScienceFiction*. Clark also produces humanist columns, essays, and podcasts for the largest dedicated website for secular discourse, OnlySky, under the title "Global Humanist Shoptalk."

Radhika Govindrajan is a cultural anthropologist based at the University of Washington who works across the fields of multispecies ethnography, environmental anthropology, the anthropology of religion, South Asian studies, and political anthropology. Govindrajan's first book, *Animal Intimacies* (2018), is an ethnography of multispecies relatedness in the Central Himalayan state of Uttarakhand in India. It was awarded the 2017 American Institute of Indian Studies Edward Cameron Dimock Prize in the Indian Humanities and the 2019 Gregory Bateson Prize, by the Society for Cultural Anthropology.

Zsuzsanna Ihar is a PhD candidate and Gates scholar at the University of Cambridge, United Kingdom. Her interdisciplinary research examines the transformation of former extraction zones, as well as the futures imagined by various stakeholders as oil reservoirs empty and the finitude of petroleum becomes increasingly palpable. She is particularly interested in the uptake of remediation, decontamination, as well as renaturalization technologies by the state, and the way in which these processes shape human and more-than-human relations. Her most recent project traces the historical convergence between militarization and agricultural development.

Noriko Ishiyama is a geographer based at the School of Political Science and Economics and Graduate School of Humanities at Meiji University, Japan. She has worked on projects focused mainly on environmental justice, American-Indian tribes, US militarism, nuclear development, and settler colonialism. Ishiyama's second book in Japanese is entitled *"Giseikuiki" no America: Kakukaihatsu to Senjuminzoku* (*America as "Sacrifice Zones": Nuclear Development and Indigenous Peoples* [2020]). It was awarded the Kawai Hayao Prize for Social Sciences and Humanities in 2021.

Eben Kirksey is an American anthropologist who specializes in science and justice. Duke University Press has published his first two books—*Freedom* in *Entangled Worlds* (2012) and *Emergent Ecologies* (2015) — as well as one edited collection: *The Multispecies Salon* (2014). Kirksey's latest book, *The Mutant Project* (2020), chronicles how profit-driven medical enterprises are pushing CRISPR gene editing into reproductive clinics.

Elizabeth Lara is a doctoral candidate at Deakin University. Her work builds on previous research about the gardens of Alcatraz Island, focusing more broadly on gardens in both current and former sites of incarceration throughout California. By examining relationships among contemporary prisons and the unfinished aftermaths of carceral heritage sites, Lara's work foregrounds common threads linking settler colonialism, migrant detention, mass incarceration, and border militarization. Her work is concerned with interfaces of carceral and abolition geographies, experiences of garden labor, social memory, and what it means to approach today's prisons as future heritage.

Jia Hui Lee is a Postdoctoral Fellow in the John B. Hurford '60 Center for the Arts and Humanities and Visiting Assistant Professor of Anthropology at Haverford College. His book project is a historically informed ethnography of human-rodent encounters in zoological research, animal training, and pest management schemes in Tanzania. Lee's research explores how more-than-human encounters in East Africa are crucial sites for generating theories and critiques that offer a counterhumanist vision of being "human" in the twenty-first century.

Kristina Lyons is an anthropologist based at the University of Pennsylvania and is also affiliated with the Penn Program in Environmental Humanities. Her current research is situated at the interfaces of socioecological conflicts, legal anthropology, and science studies in Colombia. Lyons's first book, *Vital Decomposition: Soil Practitioners and Life Politics* (Duke University Press, 2020), is an ethnography of human-soil relations that points to alternative frameworks for living and dying under conditions of armed conflict. It was awarded honorable mention for the 2021 Bryce Wood Book Award of the Latin American Studies Association. For more information, please visit www.kristinalyons.com/.

Michael Marder is Ikerbasque Research Professor in the Department of Philosophy at the University of the Basque Country (UPV-EHU), Vitoria-Gasteiz, Spain. His writings span the fields of phenomenology, political thought, and environmental philosophy. He is the author of numerous scientific articles and eighteen monographs, including *Plant-Thinking* (2013); *The Philosopher's Plant* (2014); *Pyropolitics* (2015), *Dust* (2016), *Energy*

Dreams (2017), *Political Categories* (2019), *Dump Philosophy* (2020), *Hegel's Energy* (2021), and *Green Mass* (2021), among others. For more information, consult his website michaelmarder.org/.

Alyssa Paredes is Assistant Professor of Anthropology at the University of Michigan, Ann Arbor. Her research concerns the human, environmental, and metabolic infrastructures of transnational trade between the Philippines and Japan. Her first book project identifies the conventions of crop science, agrochemical regulation, market segmentation techniques, and food standards as arenas where actors contend over the commodity chain's production calculus. Paredes's work has appeared in *Ethnos*, *Journal of Political Ecology*, and *Feral Atlas: The More-than-Human Anthropocene,* as well as in the Japanese-language volume 甘いバナナの苦い現実 (*The Bitter Reality of Sweet Bananas*).

Craig Santos Perez is a native Chamoru (Chamorro) from the Pacific Island of Guåhan (Guam). He is the author of five books of poetry and the coeditor of five anthologies. He has received fellowships from the Ford Foundation, the Lannan Foundation, the Mellon Foundation, and the American Council of Learned Societies. He is a professor in the English department at the University of Hawaiʻi at Mānoa and an affiliate faculty with the Center for Pacific Islands Studies and the Indigenous Politics Program.

Kim TallBear, Professor, Faculty of Native Studies, University of Alberta, is Canada Research Chair in Indigenous Peoples, Technoscience and Society. TallBear is the author of *Native American DNA: Tribal Belonging and the False Promise of Genetic Science.* Building on her research on the role of technoscience in settler colonialism, TallBear also studies the colonization of Indigenous sexuality. She is a regular panelist on the weekly podcast, *Media Indigena.* TallBear is a citizen of the Sisseton-Wahpeton Oyate in South Dakota.

Index

abolition: geographies, 104, 110; multi-species justice and, 5, 105–6, 110, 119, 122n34; prison, 3, 118–20, 121n6
activism, 4, 82–99; animal rights, 49n6; environmental, 6, 63–64, 75n23, 98; judicial, 69, 235; rhetoric of, 99; strategies of, 80
aerial pesticide spraying, 78, 82–88, 93, 99, 100–101n26; ban on, 95–97
Africanfuturism, 181. *See also* technology (African)
agriculture, 60; industrial, 87, 223n20; plantation, 83; plough, 36; rodent pests and, 162
Ahmed, Sara, 6, 14
Amazon Basin, 10, 54–74
Andean bear, 60
animality, 82, 89
animals, 5, 132; communication with, 215; dairy production and, 48–49nn3–4; displaced, 236; empathy with, 190–91; factory-farmed, 96; feelings of, 4; as implicated actants, 7; livestock, 36; poisoning of, 83–86, 169–71; prisoner rehabilitation and, 106; prison socioecologies and, 107; proximity to, 9; as rights-bearing subjects, 8; as sentient beings, 60; as sentinels, 16, 85; unhappy spirits of, 40; wild urban, 210, 212, 216, 218–19. *See also* Andean bear; beetles; brown tree snakes; cattle; chickens; dogs; jaguars; pests; salmon; rodents; "Supreme Court of Animals"; turkeys
Anthropocene, 9, 187, 206, 215, 219
anthropocentrism, 80, 92
Arendt, Hannah, 20n90, 126–28
articulation theory, 15–16; arranging life, 209–10; articulating love, 229; illicit technical articulations 221; microbiopolitics and, 25
Azerbaijan, 2, 10, 13, 206–24. *See also* Baku

bacteria, 2, 12, 24
Bad Actor Chemicals, 13, 87. *See also* fungicides
Baku (Azerbaijan), 13, 206–24; feral dogs in, 208–10; 214–19
bananas, 88, 163; Cavendish, 81, 83, 87. *See also* Chiquita bananas; Del Monte bananas; Dole bananas; United Fruit Company

becoming, 131–32, 189. *See also* co-becoming
beetles, 2; Coconut Rhinoceros Beetle, 84
Benjamin, Ruha, 13, 120
Bennett, Jane, 12, 224n42
Bennett, Joshua, 11, 99, 172
biocultural rights, 59, 67, 69
biopolitics, 81. *See also* microbiopolitics
blackness, 8, 98
Bogotá (Colombia) 7, 54–56; Supreme Court of Justice in, 54, 58; Tribunal of, 66, 69
Boisseron, Bénédicte, 9, 98, 108
Brown, Michelle, 113
brown tree snakes, 140–41, 145, 149–50, 154

cacao, 81, 85
California: coronavirus in, 107, 109, 116–17; penal code, 104; prisons, 3, 106, 108, 111, 113, 117–18, 120
California Department of Corrections and Rehabilitation, 116, 123n53, 124n65
camelthorn, 210, 220–21, 236
campesinos. *See* Colombia
cancer, 94, 193, 197
capitalism, 5, 92, 110, 121n11, 221; global, 10; racial, 105
carceral state, 104, 107
care, 2, 16, 113; acts of, 217; burdens of, 36–37; endangered species and, 153; justice and, 112, 235; killing and, 234; lack of, 48n4; multispecies worlds, and 209, 217; palliative, 224n39; prison and, 115–16, 120; toxic lands and, 190, 194–95, 211
cattle, 37; cow protection in India, 44, 49nn5–6, 51n30; ghosts of 38–48; as Hindu deities, 51n31; ranchers, 55, 71, 73, 196
Chao, Sophie, 8, 46, 80
chemicals, 16; altered-life and, 219–20; cancer and, 87, 94; colonialism and, 91, 201, 220; DBCP, 79, 95; DDT, 79; drift of, 78; environmental justice and, 4; food and, 168; glyphosate, 66, 79; intra-actions of, 88; lethal, 79; pesticides, 78–99, 168–71; regulation of, 94; relations of, 94; synthetic, 87; toxicity and, 24, 94, 101n28, 107, 112, 197–201, 207, 219, 229–30. *See also* aerial pesticide spraying; Bad Actor Chemicals; contamination; fungicides; hazardous waste; molecular; pesticides; pollution; rodenticides; toxicity; warfarin
chickens, 85–86, 97; dogs and, 217; frozen, 168
Chiquita bananas, 83
Choy, Timothy, 86, 109
climate change. *See* global warming
co-becoming, 13, 15, 108, 187–89, 192–95
coca plants, 57, 71; illegality of 62, 73; glyphosate and, 66
coconut, 81; farmers, 82, 84; trees, 84
Colombia, 58, 69; campesino action councils, 55; campesino communities in, 60, 68, 70, 73; environmental activists in, 63, 75n23; Indigenous groups in, 70; justice in, 7, 57; páramo, 6, 60; rights-of-nature cases in, 59–60; war on drugs and, 62, 66. *See also* Amazon Basin; Puerto Guzmán; Revolutionary Armed Forces of Colombia-People's Army
colonialism, 9; in Africa, 160; chemicals and 91, 201, 220; judicial rulings and, 61–62; in Latin America, 62; legacies of, 2, 82, 98, 105; multispecies justice and, 184; multispecies studies and, 121n11; pets and, 176n5; science fiction and, 181; in the United States, 186, 199–200; war on drugs and, 62; in West Papua, 9
compost, 106, 108, 212
conflict, 7, 234; bettering of, 56–57, 68, 72
conservation, 159; environmental justice and, 17n4; extractive industry and, 3, 66, 68; Indigenous groups and, 92; militarization of, 7, 57, 63–64, 72; peasant livelihoods and, 55, 67; rights-of-nature and, 60; Traditional Chinese Medicine and, 231
conspiracy, 86; plants and, 109

contamination: radioactive, 194, 201; water, 24, 67. *See also* pollution, chemical
coronavirus pandemic, 104, 107–10; California prison system and, 116–17, 120; incarcerated people and, 122n25; lockdown, 115
cosmopolitics, 11, 56–57, 71; between worlds, 211; conflicts and, 234; copresence and, 56; intersectionality and, 11; justice and, 25, 173; law and, 71
cows. *See* cattle

Dar es Salaam (Tanzania), 158
Davis, Janae, 12, 112, 122n34
deforestation, 55, 57–60, 62–74
deindustrialization, 118, 207
Del Monte bananas, 83
Derrida, Jacques, 15, 133, 177n19
discrimination, 9, 24–25, 125
dogs, 8, 13, 41, 85; purebred, 212; stray, 2, 208–10, 214–19
Dole bananas, 83
double death, 79, 171
Douglass, Frederick, 99, 236
Duterte, Rodrigo, 93

Earth Jurisprudence, 59–60
ecologies, 229–30; carceral, 106–7; emergent, 164; industrial and brownfield, 219; páramo, 6, 60; pests and, 176n5; plantation, 121n11
ecosystems, 5, 7, 63, 73, 83, 234; animals and, 159; protection of, 57; rights of, 6, 8, 60
Ecuador, 6–7, 59
Eglash, Ron, 161, 167
entanglement, 13; co-becomings and, 187; colonialism and, 78, 91; cosmopolitics and, 234; death and, 171; ghosts and, 46; Eva Haifa Giraud and, 80; justice and, 23, 211; oppression and, 9, 108
epistemology: feminist standpoint, 20n75
ethnography, 7–8, 236; chemoethnography, 86; multispecies, 1, 12–13; of power, 81
exclusion, 7, 13, 71; capitalist, 217; crude oil, 62; entanglement and, 234; epistemic and ontological, 18; ethics of, 16, 80, 89, 92; extraction, 3, 72, 222; extractive industries, 65, 68, 229, fugitivity and, 236; interspecies conflicts and, 207, 210; justice and, 25; politics of, 98; resource, 223n20; social, 70; techniques of 81; violence of, 209

Fanon, Frantz, 181
FARC. *See* Revolutionary Armed Forces of Colombia-People's Army (FARC-EP)
fungi, 1, 24, 82, 87. *See also* Sigatoka fungus
fungicides, 11, 78, 81, 83–84; chemical cocktails and, 86–88; health risks of, 96; names of, 87, 101n29. *See also* Bad Actor Chemicals

gaps, 184, 211; interspecies, 235
gardens, 108–9; community, 212; guerrilla, 221–22; home, 78, 81, 85, 162–63; kitchen, 106, 122n18; multispecies justice and, 112; prison, 5, 109–13, 115–16, 120–21, 122n34
Gay, Ross, 110
Gell, Alfred, 162, 164, 167, 174, 176n12
geographies, 3, 186, 188; abolition, 104, 110; carceral, 104; local, 196
Giay, Benny, 15, 20n90
Gilmore, Ruth Wilson, 104, 108, 120, 124n68
Gireau, Eva Haifa, 80, 234
global warming, 14, 17n4, 57–58, 60–62, 69, 73; climate jurisprudence, 56; climate justice, 14, 23; colonialism and, 187; multispecies futures and, 238n13; planetary histories and, 137n19
Gómez-Barris, Macarena, 11, 17n4
Goshute Indians, 196–99, 201. *See also* Skull Valley
Govindrajan, Radhika, 5, 177n26
Guam, 230
Gumbs, Alexis Pauline, 119–20

habitability, 89, 168
hackers, 174
Hall, Stuart, 15–16

Hanford Nuclear Site (Washington state), 186, 190–94, 201. *See also* Goshute Indians

Haraway, Donna, 92; *Companion Species Manifesto*, 208; on killing, 162–63; multispecies justice and, 2; on multispecies worlds, 210; on nature, 7

harm, 71, 86, 94, 99; bodily, 235; environmental, 72; responses to, 119; toxic, 201

hazardous waste, 24; dumps, 199

Hegel, Georg Wilhelm Friedrich, 129–31, 135

Hindu supremacy, 37, 49, 51n31

hospicing: capitalist extraction and, 217–18, 224n39

hospitality, 10, 92

human, the: dehumanization and, 89, 98–99, 172; justice and, 3, 24; the nonhuman and, 11, 80; as a rights-bearing subject, 8–9

human rights, 8–9; abuses, 26; animalization and, 9; conservation and, 64; environmental crimes and, 71; local idioms of, 177n28; rights-of-nature and, 59–60

implicated actants, 7, 56, 74

India, 5; cow protection in, 49n5; endosulfan ban in, 79; leopards killing cows in, 50n8; Union Carbide pesticide plant in, 20n75. *See also* Hindu supremacy; jagar; nyaya; Uttarakhand

indigeneity: animalization and, 9, 89; apocalypse and, 14–15, 61–62; bad relations and, 200–201; carceral justice and, 108; coalitions with, 3, 10, 54, 67, 73, 92; co-becoming and, 189–93; colonialism and, 55, 62, 186–88, 199–200; exclusion and, 66–68, 83; futurity and, 14–15; generative justice and, 24; genocide and, 8, 62, 191; Indigenous standpoints, 1, 188; law and, 71–72, 186; migrant work and, 30; multispecies autonomy and, 11, 70, 190–93; multispecies studies and, 121; nuclear waste and, 186–93; oppositional politics and, 54; relational frameworks and, 189; rights-of-nature and, 59–60; sovereignty and, 5, 188, 192–93

injustice, 1–6, 133; climate, 14; distribution justice and, 222n3; extinction and, 134; ghosts and, 45–47; relational care and, 209; science fiction and, 180; species of, 231; Western histories of, 181

intercultural pacts, 66–67, 71

intersectionality, 2–3, 6, 91, 230–31; climate justice and, 23; interspecies, 9–11; prison abolition and, 105–6

interspecies alliances, 9, 11, 108

interspecies relations, 162, 209

Ishiyama, Noriko, 13, 188, 191

jagar (ritual beseeching), 39, 46–48, 50nn14–15

jaguars, 55, 57, 71, 73

judicial activism, 69

jurisprudence, 58, 231; climate, 56, 61; climate change, 61, 69; Indigenous, 62, 66, 71–72; Western-colonial, 236. *See also* Earth Jurisprudence

justice: 3–6, 23–26; carceral, 23, 105; climate, 14, 23; competitive, 23, 99; contractarian, 18n28; distributive, 3, 5, 24–25, 161; ecological, 24, 80; egalitarian, 18n26; environmental, 4, 6, 10, 17n4, 23, 186, 198, 210; generative, 4, 24, 161–62, 172; limits of, 39, 181; multiworld, 4, 24, 135; procedural, 25, 177n32; racial, 12, 23, 25; recognition, 5, 25; rehabilitative, 180–81; reparative, 60, 72; restorative, 3, 25, 161, 180, 182; small, 4, 25, 211; social, 2–3, 10, 20n75, 23, 26, 112, 161; spectral, 4, 16, 26, 39–40, 45, 46, 48; substantive, 5, 26, 38; transformative, 3, 5, 26, 27n18, 72; transitional, 4, 7, 26, 57, 60; utilitarian, 18n27. *See also* injustice; multispecies justice; nyaya

Kant, Immanuel, 84, 127–30

Kim, Claire Jean, 6, 9

land, 186, 218; borderlands, 230; claimed by invasive plants, 220; colonial dispossession and, 62, 66, 83–84, 133, 186–89, 192, 194, 217; disputes over, 41, 64, 221; jurisdiction and, 230–31, languages of, 237; parklands, 214–15; porous boundaries of, 78, 81; sacred, 230; traditional rights, 188–89, 191, 198, 200; wastelands, 113, 198, 207, 214, 229. *See also* plantations; refugees; shadowlands

landscapes: active, 87, 187; Anthropocene, 215; blasted, 211, 223n20; carceral 106, 112; co-becoming with, 193–99; more-than-human, 187–93; multispecies, 211–12, 219; nuclear, 185–201, toxic, 107, 196–97, 201, 206

Langwick, Stacey, 89, 168

Lara, Elizabeth, 5, 236

law, 15, 70, 128, 137n12; British common, 136n7; carceral justice and, 23; criminal and public, 61; due process of, 93; environmental, 59, 72–73; equality before, 26; equal protection of, 96; force of, 130; justice and, 3; and order, 56, 71; personhood before, 5, 8; rights for "nature" and, 58–59, rule of, 129

legal personhood, 5, 61; of rivers, 59

legal rulings, 7; alternate possibilities for, 71; Colombia's Sentence 43, 60, 54–62, 64, 66–74; limits of, 74

Le Guin, Ursula K., 182

litigation, 68; against pesticide exposure, 95; prison abolition and, 117–18

logging: of Amazon basin, 64, 69, 71, 73

Luursema, Jan-Maarten, 232–33

Luther, Martin, 127–28, 130, 136n5

Manhattan Project, 186, 192

Mbembe, Achille, 110, 160

metaphor: animal, 86, 91; familial, 200; gardening as, 108, 115–16

microbes, 1, 107, 116

microbiopolitics, 25

Micronesian kingfisher (*sihek* or *Halcyon cinnamomina cinnamomina*), 3, 14, 140, 143, 153, 230

militarism: colonialism and, 62; conservation and, 57, 64, 71, 74; multispecies justice and, 105; colonialism and, 62;

science fiction and, 183; sentinel species and, 85; toxic legacy of, 186, 192, 196–98, 219; War on Drugs and, 55; in the Pacific Ocean, 140, 142, 150

Mindanao (Philippines), 78, 81–83, 98

mining, 55, 66–67; corporations, 74; illegal, 62, 64; industrial, 60

modernity: biopolitics of, 80; colonialism and, 14, 137n19; dividing nature/culture, 72; impurity and, 92; pesticide management and, 78, 94; plantation agriculture and, 80, 162; prisons and, 106; promise of, 131; rejection of, 224n39; trapping techniques and, 164

molecular: sovereignty, 89, 92, 94–95; intra-actions, 88; privacy, 89; relations, 94; rogue molecules, 196; species, 12

more-than-human, 89; bodily reasoning, 86; claims to integrity, 98; collectives, 85; communities, 162; entanglements, 80; inhabitants, 192, 221; material co-becoming, 194; relations, 186–87, 189, 191, 201; well-being, 174; worlds, 12. *See also* animals; chemicals; fungi; microbes; plants; stones; water

Morgan, Lewis Henry, 162, 176n12, 177n26

moringa, 85

Morogoro (Tanzania), 158–59, 162–63, 166, 168, 170–75

multinational corporations, 66, 81, 83

multispecies ethnography, 13

multispecies justice, 2–4, 17, 23, 99, 105, 236; alliances and, 11; anthropocentrism and, 80, 98; Ruha Benjamin on, 13; courtrooms and, 6; gardening and, 112; genealogy of, 17n4; growing, 236; inhuman perception and, 237n2; intersectionality and, 9–11; knowledge and, 237; language and, 234, 237n5; multiworld justice and, 135; pests and, 159–60, 169, 173–74; post-extractive, 217; prison abolition and, 105–6, 110, 119, 122n34; slime mold and, 233, 237n5; speculative fiction and, 184; visual culture and, 238n13

multispecies mediations, 207, 210, 219–20, 222

multispecies relations: breaks in, 210; cosmopolitics and, 10; decolonizing, 3; delight and, 174; human vulnerability and, 160; intimacy and knowledge production in, 177n26; justice and, 7; violence and, 168
multispecies studies, 1, 4
multiworld justice. *See* justice, multiworld
Mutongi, Kenda, 160, 176n7

NAFTA, 30
nature. *See* rights of nature
nonhumans: carceral logics and, 106; colonialism and, 78; compared with "nonwhites," 2; hierarchy of worth and, 98–99; killability of, 80; racism and, 172; symbolic violence and, 9
nuclear weapons, 186, 193–95
nyaya (relational justice), 38–40, 45

Okorafor, Nnedi, 181
olive trees, 213–14, 224nn28–29
ontology, 12, 16; multispecies, 11

Paredes, Alyssa, 5, 11
pathogens, 87, 91–92; plantation, 86
pesticides, 78–79, 87–89, 91, 101n28; Bhopal disaster and, 75, 96; regulation of, 95–98
pestiferous, the, 91, 98
pests, 99, 106, 160, 176n5; activist rhetoric and, 11, 78–80, 82, 91, 99; Coconut Rhinoceros Beetle, 84; fungal, 78, 84, 87; Integrated Pest Management, 89; plantations and, 88–89; rodents as, 159, 164, 166, 168, 172. *See also* waharibifu
Philippines, 16, 83–97; banana plantations in, 5, 11, 13; Supreme Court of, 78, 81, 95–96, 98. *See also* Mindanao; "Supreme Court of Animals"
philosophy, 3, 126; Hegel's, 135; Heidegger's, 132–33; of law, 59; Western continental, 4–5, 15. *See also* epistemology; justice; ontology
plantations, 3, 78–95; borders of, 85–86, 92; giving thanks to, 29–31; logics of, 99; pesticides and, 13, 87–89, 92–95; violence of, 11. *See also* bananas

plants, 220; beauty and, 2–3; empathy with, 190; flourishing amid hostility, 236; generative justice and, 24; as implicated actants, 7; medicinal, 109; noticing, 2; pathogens of, 87; prison gardens and, 109–10, 112, 115, 120; prisoner rehabilitation and, 106; rights of, 8; as sentinels, 16, 84–85, 99; solidarity and, 86; stubborn, 220; toxins and, 83–86, 105, 169, 171, 219–20; transnational nurseries, 214; urban (in Baku), 210, 212, 218–19, 221; vulnerable, 11; weedy, 2, 109, 220. *See also* bananas; cacao; camelthorn; coca plants; coconut trees; moringa; olive trees
pluralism: legal, 73; ontological, 12
plutonium, 196; production, 186, 193
pollution, 197; air, 24; chemical, 80, 86; environmental, 85, 107; radioactive, 193. *See also* aerial pesticide spraying; Bad Actor Chemicals; chemicals; contamination; fungicides; hazardous waste; molecular; pesticides; rodenticides; toxicity; warfarin
prisons, 106–9, 112, 123n45, 124n62; abolition and, 105; California, 3, 104, 118, 120; coronavirus and, 113, 116–17, 122n25. *See also* gardens
privilege, 9, 91, 189
protest: movements, 3; rhetoric of, 79, 80, 82
Puerto Guzmán (Colombia), 55, 62–64

race, 8, 91, 105, 121n11, 211
racism, 2, 5, 108, 220; in Africa, 160; anti-racism, 98; environmental, 4
reparations, 26, 57, 71
Revolutionary Armed Forces of Colombia-People's Army (FARC-EP), 57, 62–63
rights of nature, 6–7; Atrato River and legal personhood, 59; Colombia's Sentence 4360, 54, 56–62, 64, 66–69; Colombia's governors and, 75n12; Ecuador and, 6, 59; first local ordinance, 59; in Pennsylvania, 59; Rights of the Amazon, 54
rivers, 8, 60, 190; Atrato River, 8, 59–60,

67; Columbia River, 190; Volga River, 215
rodenticides, 158, 168–69, 171–72, 177n32; multispecies justice and, 169
rodents, 2, 158–73, 175, 176n2; *Cricetomys* sp., 165; *Rhabdomys pumilio*, 165. *See also* trapping; traps
Rose, Deborah Bird, 79, 171, 177n35

salmon, 190–93
San Quentin State Prison (California), 110–11; coronavirus and, 116–17, 123n62
science fiction, 3, 180–84, 230. *See also* speculative fiction
sentinels (sentinel species), 16, 85–86; animal, 82, 85, 99; lichen, 85; plant, 85
Sigatoka fungus, 78, 84, 87–88; control of, 100n26, 101n29
Skull Valley (Utah), 187, 196
slavery, 11, 105, 236
social movements, 2, 10, 57, 105
soils, 15, 60; Amazonian, 68; caring for, 112; toxic, 212; use of, 65–66
sovereignty, 188–89, 191, 218; cultural, 186, 188; Indigenous, 5; molecular, 16, 89, 92, 94–95; tribal, 200
species: alliances among, 3, 9–11, 108; charismatic, 214; chemical, 12–13, 219; companion, 2, 34, 106; concept of, 12–13, 129, 135, 190–91, 209; differences among, 5; empathy among, 5, 24; endangered, 14, 60, 140–56, 231; extinct, 183; feral, 218, 221; hierarchy of, 96; introduced, 213; native, 182; rights of, 6; sentinel, 16, 85; undesirable, 6; worlds of, 134–35
speculative fiction, 184, 231; Nigerian, 181
Stengers, Isabelle, 7, 10, 56, 178n40, 211, 223n22, 234–35
stones, 13
"Supreme Court of Animals" (Dagohoy), 96, 97
symbiotic agreements, 10, 234

tactics: diversity of, 108; mediations and, 211; militarized conservation, 64; for multispecies justice, 4, 18n13; psychological and emotional, 94
TallBear, Kim, 1, 13, 121n11, 188–89
Tanzania, 16, 159–75, 176n2. *See also* Dar es Salaam; Morogoro; ubunifu; waharibifu
technology, 14, 172–73; African, 160, 174; jugaad, 178n41; science and technology studies, 94
toxicity, 89, 94, 201. *See also* aerial pesticide spraying; Bad Actor Chemicals; chemicals; contamination; fungicides; hazardous waste; molecular; pesticides; pollution; rodenticides; warfarin
trapping, 159–75; anthropological literature on, 164; live, 158, 171; rodent, 162–63
traps, 160, 168; humane, 167, 176n5; rodent, 16, 158–75; for stray dogs, 216
trauma, 184, 194, 218; coronavirus pandemic and, 108, 120; humane traps and, 176n5; multispecies, 184
Tsing, Anna, 2, 211, 223n20
turkeys, 30–31

ubunifu (innovation), 161, 166, 174
US Department of Energy (DOE), 192, 195, 201
Uttarakhand (India), 34, 36, 49n6

violence, 8; anti-Black, 104; capitalist, 222; in Colombia, 57, 60, 62, 70; against cows, 36, 48n3; against Dalits, 37, 49nn5–6; depictions of, 218; Enlightenment thinking and, 224n39; environmental, 193; of extraction, 209; to Indigenous peoples, 187; interspecies, 207, 217; justice and, 26, 127–29; kinship and, 99; against Muslims, 37, 49nn5–7; ontological, 1, 56; prison gardens and, 112–13; purity and, 92; racial, 25, 118, 219; racist, 91, 172; rodent traps and, 162; shared experiences of, 3; of slavery, 236; slow, 8; state, 3, 106; structural, 14, 55, 74; symbolic, 9, 122
vulnerability, 119; ecological, 5; global warming and, 14; justice and, 18n25; pollution and, 24, 85; racism and, 108

waharibifu (pests), 160, 163
Wanapum Indians, 190, 192, 194, 201
warfarin, 168–69, 172, 177n32
water, 13, 88–89, 190–92; access to, 109; camelthorn and, 220; chemicals in, 168–69; co-becoming and, 187; cocktails, 93; contamination, 24, 193; Masan (deity) and, 43; pipelines, 11, 215; quality, 112; rights to, 58; sources, 66, 78; warfarin in, 177n32
weeds, 106, 108–9, 120, 210, 212, 215, 221, 230
West Papua (Indonesia), 8–11, 14–15
Whyte, Kyle Powys, 14, 187